The Complete
GRANNY MILLER

A Mind to Homestead
&
Garden Farming

Katherine Grossman

ISBN: 978-1-7321663-3-2
Revised Edition - August 2023

GRANNY MILLER PRESS
PO Box 116
Harrisville, PA 16038 -9800

Author website
www.grannymiller.com

This book is dedicated to Albert H. Grossman
1919 - 1992
With lasting love and gratitude

❦

And to Glenn P. Grossman
1929 – 2001

You were right about the house.
We should have pushed the damn thing over.
You were right about the trees, too.

❦

And this book is dedicated most of all to my beloved husband.
Without him, I would never have written a word of it.

Richard L. Grossman
1963 – 2020

I'd do it all again.
You know I would.

Contents

About This Book

A Word from the Author

I'd like to thank you for your purchase of THE COMPLETE
GRANNY MILLER. This work is a compilation of my two
previous books, A MIND TO HOMESTEAD and GARDEN
FARMING, in a single format.
Both books were a collection of selected articles and posts from
my old and long-since-abandoned website, GRANNY MILLER
(2006 – 2016). Both books are out of print and are unavailable as
electronic media.
I decided to republish without the black and white photos that had
been published previously because I was never pleased with how
the photos appeared in either of the two books.

In the past, my goal for publishing both books was to share
the information they contain at an affordable price. That goal
remains the same.

For the most part, the content of both books has remained
essentially unchanged. I corrected and edited some poor grammar
and sloppy spelling, which I'm infamous for. Where I could, I tried
to remove my tendency to pomposity and long-windedness. In
some places, I just couldn't manage it without rewriting entire
sections. I apologize for that.

I hope that you, the reader, will find something within these pages
that will be of benefit to you and your life. It has been my pleasure
to make some of the old GrannyMiller website content available
once again.
I hope you'll enjoy it.

Katherine Grossman
Pennsylvania
May 2020

BOOK 1
A MIND TO HOMESTEAD

Introduction

How The Whole Thing Got Started

Halloween is a wasted effort at my house. But that never stops me. Every year without fail, I load up on candy in anticipation that after dark, children will show up at my front door dressed in spooky homemade costumes. Night falls. I wait.

And not one trick-or-treater, large or small, knocks at my door. Typically, I give up hope around 8 o'clock and turn off the porch light. And then eat all the candy.

It's been that way for over 20 years.

But the night of Halloween 2006 was different. On a sheer whim while surfing the Internet and peeling wrappers off of Fun Size Hershey's candy bars, I began a Google Blogspot blog while blitzed on sugar.

The notion of being able to upload images and uncensored content to the Internet for free, with the possibility of a worldwide audience, intrigued me. I called my new blog GRANNY MILLER. I took the idea and concept for Granny Miller from Mother Jones and Carrie Nation. I wanted to convey the sense of an older and mature agrarian social activist, so I chose 'Granny' as my feminine description. I picked the name 'Miller' because I've been known to get carried away on a political rant, and I free-associated the concepts of an 'ax to grind' and 'grist mill'. The possibilities at the moment seemed endless.

Granny Miller was going to be my platform for social and political change. In my mind, Granny Miller's mission was heroic and epic. I was determined to be more than a digital Don Quixote. Granny's purpose was to sow the seeds of agrarian anarchy and to spread the gospel of personal self-reliance around the world.

It seemed to me that a complacent and somnolent American public was in danger from large multinational corporations and Big Bad Agriculture. People were in need of an awakening - and fast! I reckoned that railing against the USDA and NAFTA, along with a few homemade how-to canning videos and pantry lists, would help me awaken the sleeping masses.

Like I said, I was totally buzzed on sugar.

I think what surprised me most about my entire grandiose Granny Miller endeavor was that I found such a ready audience. I just couldn't believe that strangers and part of the establishment media actually took me seriously. I was astonished! Young people, especially, were hungry for the simplest and most mundane household and small farm information. My soapbox worked out well for a while until I grew weary from sitting at a keyboard. I decided that real life was more exciting and worthwhile than my digital and screen life, so I deleted Granny Miller from the Internet. In fact, I deleted Granny Miller not just once, but at least five times between the years of 2006 - 2016. Sometimes, being epic and heroic can be too time-consuming.

When I first started Granny Miller, there weren't many small farm or homesteading blogs. Back then, reliable information on homesteading and traditional agrarian life was scarce and hard to come by. But things have changed a lot since then. These days, there are numerous first-rate small farm and homesteading websites and blogs. Self-published print books and eBooks about homesteading and related topics are easily found. Even big dinosaur print media and the entertainment industry have jumped on the self-reliance and simple living bandwagon.

There has been a renaissance in garden farming and smallholding across the country. Ordinary farmers have become mythologized folk heroes and marketing gurus. Home canning has made a comeback, and well-stocked pantries and root cellars are a source of household pride and security once more. Backyard bees, small stock, and chicken keeping are completely mainstream again. So much so that many cities and municipalities are routinely being asked to change local zoning ordinances to accommodate poultry and small livestock keepers. Local farmers' markets are ubiquitous in almost every town and city across the country. Even vegetable gardening has returned to the White House lawn. Everywhere, young (and not so young) people are questioning the value of a life driven and fueled by overconsumption, consumer debt, and mindless materialism. Many people are seeking and envisioning a fresh and more authentic agrarian way of life. But often that new life is a life that they aren't prepared for. Frequently, an entire new skill set and altered worldview are required in order to live in the way that they've dreamed about.

One thing that sets Granny Miller apart from other homesteading or garden farm websites is that I'm not a neophyte. I'm an ordinary Pennsylvania farmwife in my mid-60s. What may be of interest to some is that up until the age of 35, I lived my entire life in large metropolitan areas. Truth be told, I'm a city transplant who grew into a country person.

What I've discovered about self-reliance, rural life, and homesteading has only been within the last thirty years or so. I get a lot of questions from people who live in the suburbs or in large cities and want a more self-reliant or simpler and quieter life, but don't know where to begin. Sometimes they express doubts as to whether or not they can learn the skills they'll need to build that life. I share this information about myself because readers often assume that I have always lived a rural life and was born knowing how to render lard or milk a goat. That's just not so.

2

Thirty years ago, I knew nothing about cows, water witching, wood stoves, guns, or so many of the things that I know about today. Some of the skills and knowledge that I now take for granted, I had to learn on my own. Often, practical experience supplemented by the local public library was invaluable to me. I had to learn many things about country life the hard way.

My country education was helped a great deal by marriage. Part of what I learned about plain living and self-reliance was passed on to me by my husband's family. In many matters, I was schooled by the two generations of rural Pennsylvania country folk and small farmers that came before me. Sadly, most of those people are now gone. But the information and traditions that they shared still live on in me.

My agrarian outlook and life have been acquired and carefully cultivated. I think it's reassuring for some people to know that it is possible to make the transition from city life to country life; from a life of total food and energy dependence to a life of relative independence. If you really make up your mind to do it, a more self-determined and self-reliant existence is possible. No matter who you are. In fact, it's more than possible.

A more sustainable and self-reliant life isn't dependent upon geographical location, education, or financial resources. In the suburbs, on a tiny town lot, or even in the big city, you can become more responsible for your own basic needs. Whether you are young or old, a man or a woman, rich or poor, it doesn't matter. Sustainability and self-reliance are really about the choices we make every day. Wherever we are and whoever we may be. With the publication of this book, I endeavor to repay what was so freely given to me. It is my sincerest wish that this book will be of some benefit and provide encouragement to homesteaders and small-holders looking for a better life.

Enjoy the trip!

Katherine Grossman
Plain Grove, Pennsylvania
Winter 2016

The Complete Granny Miller

Chapter 1

"You will never do anything in this world without courage. It is the greatest quality of the mind next to honor." – Aristotle.

It's A Hard Life, But It's The Good Life

The following short essay was first published in December 2012 on my website GRANNY MILLER. It later appeared in print in LIVING READY MAGAZINE. Unfortunately, when it did appear in print, it was at the back of the magazine. It had been edited and illustrated with images that I thought were inappropriate for the subject matter.

Numerous women and some men have taken extreme exception to several of the ideas and opinions put forth in my essay. But happily, just as many men and women have found hope, comfort, and validation from it. For the past four years, I have been endlessly entertained by the comments and speculations I have seen online about its author (myself). If they only knew! On occasion, I've been asked if, given the chance to do over, whether or not I'd write the essay the same. I have to say, "I'm not sure, honestly." That said, I still maintain all the basic suppositions of the essay.

In a society where gross consumerism is considered a national virtue instead of a national vice, it's to be expected that the traditional economic union and division of labor between women and men is actively discouraged.

Survivalist Prepper or Housewife?

Lately, my husband and I have been having an ongoing conversation about the so-called 'survival' and 'prepper' movement. We are both more than a little dismayed with the entire phenomenon. Many people seem to misunderstand the concepts, skills, and life choices often associated with prepping or survivalism, regardless of the name. Frankly, I find the terms survivalism or prepping to be positively inane. From my point of view, much of the prepper and survivalist mentality seems born of an apocalyptic Hollywood and TV Land fantasy founded upon fear, ignorance, and unabashed consumerism. Many of the skills and life choices often associated with modern survivalist living or prepping were, at one time, the skills and choices of simple living and traditional old-fashioned common sense.

So, you can imagine my chagrin when, 15 years ago, a dear friend described my life as that of a survivalist. Her interpretation of my life was a backhanded compliment. I understood what my friend was trying to convey. She was trying to find the words to communicate and illustrate a lifestyle that is self-sustaining, non-consumer-oriented, and not wholly dependent upon the grid. However, her comment gave me food for thought. I found it interesting that in less than two generations, the average American household had become, for the most part, an isolated and non-productive, debt-driven economic model. The concept of a traditional and contained household economy had become alien in late 20th-century America. Even more disconcerting, the traditional American housewife had become a rare thing. She was going the way of the dodo bird.

The most basic of human needs revolve around food, clothing, and shelter. Those needs can only be met by a realistic understanding of who we are as people. That understanding, combined with a consideration of our particular environment, and with a sensible stewardship of our circumstances, permits fundamental life needs to be met. Those are some of the exact ideas and concepts that preppers and survivalists have recently begun to realize. And that's a really good thing.

The lack of traditional housewifery and the lost concept of self-contained household economies, which are interconnected with other traditional households, seem in part to have been the breeding ground for the modern survivalist and prepper moment. America became a helpless and dependent society the day American housewives quit working for themselves and their households, and became wage slaves for someone else. For most of the 20th century, the American housewife didn't work for wages. Food, clothing, and shelter were her specialties. Whether she lived in a city or the country, she knew how to cook, sew family clothing, kill and dress a chicken, and get by without electricity or indoor plumbing if she had to. She had a full pantry and a backyard garden. She raised her children. She had plenty of time for outside interests and her community. She also didn't have a car payment, a TV, or credit card debt. She managed to marry and stay married. The average American housewife from 1920 to 1970 would today be considered a survivalist. For many of you reading this, your great-grandma was a hard-core prepper.

Today, most American households are non-producing households. The concept of a productive domestic economy has long been forgotten, along with the skills needed to sustain that household economy. Nowadays, the average American household is a consumer-based closed-system economy. Just about everything - from food to entertainment to underwear - is produced outside the confines of the home. Most adults work to make money so that they can pay someone else to make or provide for their basic material and non-material life needs. Many American households actually produce little for themselves except debt, depression, divorce, and bratty kids.

The notion of a self-contained household fueled by self-labor, traditional sex roles, and frugality was the exception years ago when my friend described how I lived my life. Hopefully for America, I think there may be the beginnings of a social change. Preppers and survivalists could be the first hints of that reawakening. For me, my life has been that of a traditional housewife and farmwife. I do not need to call it anything else. The fundamentals of what I have done with my life and the way that I have consciously chosen to live my life could have been achieved while living in the suburbs of Nashville or the Upper West Side of Manhattan.

5

While I could not have raised cows or sheep on a 100' X 60' town lot, I most certainly could provide for my household by sewing, gardening, canning, and raising small livestock (in my basement or garage if I had to). The fact is that an off-grid source or backup for water, heat, lighting, and toilet needs is possible, no matter where one lives. It takes no great skill to refuse consumer debt and to use only cash. To live beneath your means is attainable and prudent for everyone. A measure of self-reliance and economic security takes hard work, self-control, and discipline. But it's within the reach of anyone.

So, the next time you see a picture of my pantry, or read how I spin yarn, or plant onions, remember I'm doing what every traditional household economy has always done. It is not exceptional. The fact of the matter is that a household that doesn't provide for its own needs is the historical exception.

I work for myself. I provide for my household. I literally make my living by the sweat of my brow and from my labor. You can do it too. I'm not a survivalist or a prepper. I'm a traditional housewife.

101 Basic Homesteading Skills

The following list was first published on December 31, 2012. My husband and I compiled this list quickly and effortlessly while sitting at the kitchen table after breakfast. I had no idea the impact it would have on my readers and those seeking a life of preparedness. Since its first publication, this list has become the most copied and plagiarized content from my website. At the time of original publication, it struck a chord with many preppers and new homesteaders.

What follows is a partial list of homesteading skills that my husband and I share between us. I offer this list as a starting point to hopefully inspire young couples or new homesteaders to learn at least one or two new skills.

Why not finally get rid of the TV this year and ditch Facebook? Dare to find a new way of life. Every skill that you learn will be one step closer to a life of greater self-reliance and independence.

1. Learn how to safely use a chainsaw
2. Learn how to grow a vegetable plant
3. Learn how to sharpen any edge tool - knife, axe, hoe, chisel, etc.
4. Learn basic firearm safety and gun-proof your children and grandchildren
5. Learn how to dub a chicken
6. Learn how to read the weather

7. Learn how to spin wool, cotton, or flax into thread or yarn on a spinning wheel or with a drop spindle
8. Learn how to use a garden shovel, spade, or hoe without hurting your back
9. Learn how to light a fire indoors or outdoors
10. Learn how to go to a country auction and not get skinned
11. Learn how to crochet
12. Learn how to butcher small livestock like rabbits or chickens
13. Learn how to hang clothes on a clothesline
14. Learn basic tractor maintenance
15. Learn the differences between trees and the unique properties of various types of wood
16. Learn how to cook 10 basic meals from scratch
17. Learn how to pasteurize milk
18. Learn how to witch for water with a forked branch or a bent metal hanger
19. Learn how to distinguish healthy plants and animals from unhealthy plants or animals
20. Learn basic sewing skills
21. Learn how to set an ear tag or tattoo for animal identification
22. Learn how to determine an animal's age by its teeth
23. Learn how to cut and glaze glass
24. Learn how to drive a standard transmission vehicle
25. Learn how to thaw out frozen pipes without busting them
26. Learn how and when to use hybrid seeds
27. Learn how to hand thresh and winnow wheat or oats and other small grains
28. Learn how to train a working cattle or sheep dog
29. Learn how to read the moon and stars
30. Learn how to make soft or hard cheeses
31. Learn how to live within your financial means
32. Learn how to fillet and clean a fish
33. Learn how to use a wash tub, hand-wringer, and washboard
34. Learn how to make soap from wood ashes and animal fat
35. Learn how to lay basic brick or build a stone wall
36. Learn basic home canning and food preservation

The Complete Granny Miller

37. Learn how to save open-pollinated seeds
38. Learn how to de-horn livestock
39. Learn how to use an awl and basic leather repair
40. Learn how to make long-term plans for the future, like planning an orchard or a livestock-breeding program
41. Learn the mental skills necessary to jury-rig anything with duct tape, baling twine, or whatever is on hand
42. Learn how to read an almanac
43. Learn how to euthanize large livestock safely
44. Learn how to cook on a cook stove
45. Learn how to entertain yourself and live without electronic and digital media
46. Learn how to shear a sheep
47. Learn how to manage human urine and feces without plumbing
48. Learn how to swap, barter, and network with like-minded people
49. Learn how to make a candle
50. Learn how to dig and properly use a shallow well
51. Learn how to refinish furniture
52. Learn how to drive a draft animal
53. Learn the mental and spiritual skills to realistically deal with life, death,and failure
54. Learn how to use non-electric lighting
55. Learn how to caponize a cockerel chicken
56. Learn how to restrain large livestock
57. Learn how to use a treadle sewing machine
58. Learn how to give an injection
59. Learn how to properly use a handsaw, hammer & nails; screwdriver, wire cutters, and a measuring tape
60. Learn how to recognize your own physical and mental skill limits
61. Learn how and when to prune grapes and fruit trees
62. Learn how to hatch out chicken, duck, or other poultry eggs
63. Learn how to use a scythe
64. Learn how to skin a furbearer and stretch the skin
65. Learn how to tell the time of day by the sun
66. Learn how to milk a goat, sheep, or cow
67. Learn how to stomach tube a newborn animal
68. Learn how to break ground and plow

69. Learn how to use a wood stove and how to bank a fire
70. Learn how to make butter
71. Learn how to knit
72. Learn how to make and use a hotbed or cold frame
73. Learn how to deliver a foal, calf, lamb, or kid goat
74. Learn how to know when winter is over
75. Learn how to plant a tree
76. Learn how to brood day-old chicks
77. Learn how to dye yarn or cloth from plants
78. Learn how to haggle like a horse trader
79. Learn how to bake bread
80. Learn how to use a pressure tank garden sprayer
81. Learn how to halter break a horse or cow
82. Learn how to graft baby animals onto a foster mother
83. Learn how to weave cloth
84. Learn how to grow everyday kitchen herbs
85. Learn how to make sausage
86. Learn how to set and bait traps for unwanted vermin and predators
87. Learn how to grind wheat into flour
88. Learn how to make paper and ink
89. Learn when it is more economical to buy something ready-made or when to make it yourself
90. Learn how to castrate livestock
91. Learn how to choose a location for a vegetable garden or orchard
92. Learn how to weave a basket
93. Learn how to use electric netting or fencing
94. Learn how to make fire starters from corncobs or pinecones
95. Learn how to use a pressure cooker
96. Learn how to correctly attach 3-point hitch implements to a tractor
97. Learn how to trim the hooves of goats or sheep
98. Learn how to sew your underwear
99. Learn how to make wine
100. Learn basic plumbing and how to sweat copper pipes and joints
101. Learn how to reload ammunition

A Morning Prayer

Often when I'm asked,
"What's the most important or necessary thing for a homesteader or small holder to possess or to own?"
I always say, "Faith in God."
If the person asking the question can't deal with the word or concept of God, I reshape my answer for that person. For them, I explain that it is necessary to have faith in a supreme order and in the rightness and goodness of creation and the universe.

My life has taught me many things. But the most important thing that I have learned, I will share with you now. It is this.
That without trust in the absolute sovereignty of God, and lacking an understanding of your particular role and place in this universe, the inevitable disappointments, heartbreaks, and sorrows that come with small farm life and homesteading will at times be unbearable. The grief and inescapable failures may break or destroy you. Many of the happiest moments of my life have been spent in my barn watching new life miraculously enter the world. And conversely, some of my most desperate hours have been in that same barn. Deprived of some sort of faith, it is almost impossible for the average person to make sense of cruel suffering, needless loss, and death.

In 21st-century America, there are no new frontiers for homesteaders or pioneers to settle and claim for their own. For the vast majority of us, reliance upon and a trust in Divine Providence is not necessary for our day-to-day survival. For most modern people, spiritual idleness and lethargy are, I believe, a consequence of a cushy and comfortable technological existence. The majority of Americans know little of the harsh realities and brutality of the natural world. But it hasn't always been that way.

The pilgrims and settlers who came to the New World at Jamestown and Plymouth managed to endure dangers, famine, hardships, and unbelievable losses, due to their faith alone. It wasn't all that long ago on the prairies of South Dakota and the farms of Minnesota that early pioneers and farmers survived the Children's Blizzard and outlasted five years of the Grasshopper Plagues. Fortified and strengthened with only personal faith and community prayer. Where there is risk and peril, there is also God.

The only new frontiers left for us are outer space and the oceans deep. So, whether you hear about it or not, I guarantee you that the first men and women to Mars and beyond will have an abiding faith in God.

The following prayer is known as The Prayer of the Optina Elders. I have loved and said this prayer at the beginning of every new day for over 40 years.Frequently, a line or a phrase from this prayer will come to me during the day. This prayer is etched upon my heart and has fortified my body and my soul through many anxious moments.

The Complete Granny Miller

A Morning Prayer

Grant unto me, O Lord, that with peace of mind I may face all
that this new day is to bring.

Grant unto me to dedicate myself completely to Thy Holy Will.

For every hour of this day, instruct and support me in all things.

Whatsoever tidings I may receive during the day, do Thou teach
me to accept tranquilly, in the firm conviction that all
eventualities fulfill Thy Holy Will.

Govern Thou my thoughts and feelings in all I do and say.

When things unforeseen occur, let me not forget that all cometh
down from Thee.

Teach me to behave sincerely and rationally toward every
member of
my family, that I may bring confusion and sorrow to none.

Bestow upon me, my Lord, strength to endure the fatigue of this
day and to bear my part in all its passing events.

Guide my will, and teach me to pray, to believe, to hope, to
suffer, to forgive, and to love.
Amen

The Complete Granny Miller

About two years after I started my website GRANNY MILLER, I decided
that it would be fun to do a monthly 'Ask Granny' feature.
The idea was that it would be a little like Dear Abby, but specifically for
homesteaders.
Here are five good questions that I received, and my responses to those
questions.

How Much Time & Land Is Needed to Survive Off theLand?

Dear Granny -
You have a lot of canned goods that you did yourself, and seem to farm
quite a bit. How much time (in hours per day on average) do you invest in
order to successfully be food self-sufficient over the course of a full year?
To go along with this question, how much land and time would you have to
have in order to survive off the land? As in pay your bills by selling food.
Kevin

Dear Kevin -
Your first question is easy to answer; the second is not so easy. It's not
possible to enjoy what is considered a modern and varied diet and be 100%
food self-sufficient. A life without sugar, coffee, olive oil, oranges, lemons,
cocoa, and other products that aren't native to Pennsylvania would be hard
for me to tolerate year in and year out. While it is possible to live off of just
the food you produce, it's a horribly monotonous diet. Much of my home
food production is seasonal, especially during the late months of summer.
However, some food production is ongoing daily throughout the year. The
home-canned food that you see on my shelves comes from a large vegetable
garden, a small orchard and grape arbor, and from assorted animals I raise.
Each season brings a different type of work.

During the summer, many hours are spent in the kitchen and
the garden. During canning season, which spans from June to
October, 12-hour days in the kitchen and garden, 2 or 3 times a
week, are not uncommon. Serious home food production is
more than a full-time job.

Throughout the year, animals require daily care. The morning
and evening chores to feed and care for my livestock take about 45 minutes
twice a day, with all day most Saturdays and Sundays devoted to barn work,
fence repairs, and animal health.
In early spring, I'm in the barn for about 4 hours every day during lambing
season, which can last 3 or 4 weeks. Washing eggs that need to be gathered
every day is another daily chore. Except for hard winter, 15 minutes each
day is spent gathering, washing,and sorting eggs.
On average, 3 hours every day of my life is spent cooking, baking, and
cleaning up the kitchen mess that comes with food self-sufficiency. And

The Complete Granny Miller

that's not including regular housework. Regular housecleaning, ironing, laundry, sewing, and other domestic household tasks still must be done. During the cold weather months, I spend about 2 hours spread throughout the day tending to a wood-fired cook stove to cook the food, and to the wood and coal stoves that heat my house.

During the spring, my husband and I spend a couple of weeks with a few 14-hour days plowing, tilling, sowing, and planting. Right now, we are making hay, which is another couple of 12-hour days. Hay making is spread out 2 or 3 times during the summer.

Just so you know, what I have illustrated above is not considered 'farming'. What I have described is the labor and effort involved to feed two middle-aged adults with no children at home, and nothing much extra to sell. We are using modern diesel engine equipment and not horsepower. If we used horses or mules, it would be less expensive but much more work. We sell some eggs, apple cider, chicken, beef, pork, and lamb. But the income is nowhere near enough to live on. It won't even pay the taxes and auto insurance.

What my husband and I do, I call a closed and sustainable domestic economy or garden farming. Some people call it homesteading. But nowadays, the new name for that kind of life is called 'prepping'. Whatever you want to call it, the arrangement works because my husband and I have embraced traditional agrarian sex roles and an old-style cottage economy. You don't need lots of land to produce some of your food. Many people can produce much of their food with a large garden, a dairy goat, backyard chickens, rabbits, and a few fruit trees. Three generations ago, that's what most people did, whether they lived in towns or cities.

Your second question cannot be answered in general terms. Making money from your land depends upon where you are located, what type of ground you are working, your markets, and how healthy and strong you are. I know of people who can make a good living on 5 acres and those who can't make it on 500.

The key to self-reliance and living off the land by your efforts is to, as much as possible, have no debt or needless bills. Own everything outright. Barter your time for goods. The powers that be have yet to find a way to tax our time or cabbages. Grow as much of your food as possible and learn to do without. People use and mindlessly purchase items every day that they don't really need to be happy or to live well. Thrift is a forgotten virtue. If I had to make a living from just farming, I don't believe it would be possible given my location and circumstances without cheap labor. The farm labor issue is an almost insurmountable problem for small family farms. Small farms are hamstrung due to overregulation. I own good, fertile farm land and two tractors. But I would need to hire at least two or three dependable people and pay them sub-minimum wage, even to stand a chance to break even by farming—the best of luck to you in the future.

How to Keep Rabbits from Eating Just A Few Vegetable Plants?

Hey Granny! –
I just started growing some of my own vegetables this summer, and they are just starting to come through. I am very proud of my two small tomatoes and three hot peppers, but I do not want rabbits to eat them. What would you recommend to keep them away?
Thanks!
Vegan Student

Hey Vegan Student,
Just like people, wild animals like vegetables. We have problems here with deer, groundhogs, and rabbits. As farmers or gardeners, we have only limited choices.

- A motion-activated pest deterrent water sprinkler
- Shoot and eat them
- Or as large commercial vegetable growers do, shoot them and leave them to rot
- Trap them in a box or leg hold traps
- Fence them out
- Attempt to use a chemical repellent

If you have access to a garden hose near your garden, consider motion-activated pest deterrent water sprinklers, which are effective. It's what I use in my kitchen garden. Many models have a height adjustment feature that allows the spray to be aimed high or low. Just be careful you don't forget when it's on and walk in front of it!
If you are a vegan, you may be opposed to hunting or trapping. In my experience, however, commercial grain, fruit, and vegetable growers actually kill more wildlife than conventional livestock farmers. You could live-trap the pest. But if you trap and move an animal, you are generally sentencing it to death anyway. The newly released animal will be perceived as either a predator or competitor and will be driven out of its new surroundings or be killed.
I've never tried chemical repellents. But from talking to people who do use chemical repellants, many don't work, and the ones that are effective stink. Some people use human hair with success. Human hair is free for the sweeping at your local barber or hair salon. Use the hair sparingly, like mulch around the plants. If hair doesn't work, your only practical choice may be to fence the plants in. A small circular fence can be held in place with one stake. You can either buy a small roll of hardware cloth or build a 2-foot-high garden fence to surround the plants. This will be inconvenient, as you will need to move the fence to weed or pick vegetables. You could spend more money and fence in a bigger area, but for just a few plants, it's not worth it.

Rabbits are generally easier to fence out than deer or groundhogs.
Good luck!
And speaking of rabbits…

What Is the Easiest Animal To Raise For Meat?

Dear Granny -
What is the easiest edible animal to raise for meat? What is the best starter animal in price and in headaches. When I get my own farm going, I hope to not have to do a lot of vet trips.
- Lee

Dear Lee -
Without a doubt, rabbits are the easiest and most cost-effective animal when it comes to putting meat on the table. Going by today's prices (2012), $185 would probably set you up with cages, water bottles, feed pans, and grade breeding stock. That would be two does and a buck: each with their cage and equipment.
Not all rabbits are suitable for meat production. Stick with the Californian, New Zealand, and Creme D'Argent breeds. Rabbits are especially intolerant of extreme heat, so you need to keep that in mind. As to having to call a veterinarian for a sick rabbit, that's probably not going to happen. The vet call will cost you more than the rabbit. The fact is, a sick rabbit is a dead rabbit.
Rabbits have historically been the animals of cottagers and homesteaders. They don't take up a lot a space and pay back in a big way.
Rabbits were one of the reasons people in the Netherlands managed to survive World War II. Rabbits are easy to keep in cities and towns. Rabbits work in perfect tandem with vegetable gardens. Rabbit manure can be applied directly to plants in the garden, and won't burn plants, and does not have to be composted or aged. Basically, you feed the rabbits from the garden, and then recycle what you fed to them back into the garden.
Rabbits are much easier to dress out than chickens. They are more efficient in turning grain, vegetables, or kitchen waste into food for humans. Rabbit meat is high in protein and low in fat; much lower in fat than chicken. Rabbit meat is good for home canning, and canned rabbit can be used exactly like canned chicken in recipes.
The hardest thing about raising rabbits for meat is killing them. It can be difficult to kill a helpless, furry, cute animal.
The best way to kill a domestic rabbit (and chickens, too) is by cervical dislocation. But not everyone is up for killing rabbits or chickens with their bare hands or a broomstick. A helpful tool for killing rabbits is the Rabbit Wringer™. It's fast and doesn't require a lot of strength. It is a bloodless method of killing, and if done correctly, it is the most humane butchering

The Complete Granny Miller

method I know of.

For some reason, rabbit meat fell out of favor in this country sometime after the 1920s. Rabbit meat is hardly ever seen in the grocery store these days. When you do find it, it can be expensive. And that's a shame because a rabbit stew made with potatoes, parsnips, and carrots makes for a delicious and nutritious supper.

Picture book farms and neat tidy homesteads only happen in the movies and on the Internet. Keep in mind that the Internet is full of deceptive, pretentious, and elusive information and images.

No one has a perfect life. Everyone of us only gets 24 hours in a day. But how you choose to use those 24 hours is what can make or break a small farm or homestead.

How Do You Keep on Top of Homestead Tasks?

Dear Granny -

Your home always looks so lovely; neat, clean, and orderly indoors and out. My question is: How do you keep on top of all the various, diverse homestead tasks?

Do you use calendars or a series of calendars, or is it all just second nature after all this time? My question is: I struggle with keeping on top of things here on just 12 acres with far fewer animals! Try as I may, I have yet to find a 'system' for staying on top of everything.

  Cynthia

Dear Cynthia

Thank you for the compliment. Actually, for the most part, I don't use any type of calendar system except for animal health matters and keeping an orchard spray schedule. A calendar is used to note when calves or lambs are due to be born.

My 'secret' for staying ahead of things is that I'm a full-time farmwife. I no longer have children at home. My home and farm are my entire working life. I don't work outside of my home, and I allow nothing else to get in the way of my own 'home economics'. I work much harder for myself than I would ever work for someone else.

"By the inch, life's a cinch. By the yard, life is hard."

At the beginning of every new week, I make a list of all the chores or projects that need to be accomplished during that week. I prioritize my list

The Complete Granny Miller

and try to be realistic about my energy level.

Anything I don't finish in one week gets rolled over to the next week's list. I work steady until my weekly goals are completed.

I dedicate two days each week to laundry and housework. Every week, I try to save one day for errands away from home or for unexpected occurrences: a down cow, a flooded basement, a morning migraine, or a visit by a home repairman.

Basically, I try to give a cushion for the unforeseen and the unplanned.

I also try to set aside one day a week just for myself to do whatever I like.

"Don't own so much clutter that you will be relieved to see your house catch a fire." - Wendell Berry.

Clutter is kept to a minimum inside and outside of my home. I pick up and clean up as I go along - all day long. That means buckets, feed pans, and baling twine when out-of-doors; and dishes, coats, and junk mail indoors. Everything has a place, and I don't allow myself the luxury of procrastination. Unless I am physically unable to continue without rest, I never leave a day's work incomplete. My home is small. If I run out of space or the clutter gets out of control, it's a sign that I have too much stuff. I routinely pitch out what I really don't need or give it away.

Please don't forget that you are only seeing selected snippets from my life. I, too, have lots of homesteading frustrations. The Internet isn't real life. You only see what I want to show and not the complete truth. The picture of my life displayed to the public has been cropped, altered, and digitally enhanced.

Homesteading means different things to different people. Sometimes the desire for a life of homesteading or small holding can become a problem between couples.

The truth is, not everyone has the same dream or the same needs. Most people would rather spend a week at the beach instead of weeding a 20 ft. row of carrots at home.

At times, homesteading can be a source of conflict in families with children, especially teenagers.

The problem of a homestead lifestyle clash is more prevalent than you might first imagine.

The Complete Granny Miller

She Doesn't Want To, But He Does

Hello Granny Miller,

I have a question for you. I was raised on a farm and lived the lifestyle of self-reliance. After a lot of years working for someone else I am trying to retire. I want to get back to that self-reliant life. I know how to garden; I have just gotten some chickens and ducks. My next is rabbits and quail. I want a bee hive or two. I live in town so I can't have bigger livestock there; I am looking into leasing some land. So as you see I think I am heading in the right direction.

Now I know all of this is a lot of work. My wife and I are both are hard workers but she thinks I am a prepper just because I want to have enough food on the shelf or in the freezer. The one thing I have her hooked on is the ducks! She loves to see them in the pond and the silly things they do (so I may have a head start). We also have wild Mallards that come and have babies at our house every year. This is how I got her on the ducks in the first place. So some words of wisdom, some direction, or any help to get her to understand the direction I'm headed would help. I know she will love the journey.

Thanks.

Stephen

Hello Stephen,

You will probably not like my answer. Many people desirous of a more self-reliant life have a partner who is not interested in gardening, small livestock, or any type of self-determined living situation. Changing and evolving lifestyle choices are frequently a serious issue between couples, especially among older couples thinking about retirement. In fact, almost 80% of all couples planning for retirement will run into some type of life choice or living situation conflict or issue.

"You can lead a horse to water, but you can't make it drink."

What I hear in your question is that your wife isn't interested in a different way of life. Maybe she doesn't appreciate or understand why you would want to change. A change in a partner can be scary and confusing. Maybe she feels like you are trying to drag her where she doesn't care to go. Or now that you are thinking about retiring, you are trying to change the rules and upset the way things have always been. I don't blame her for feeling less than enthusiastic.

Maybe she feels like you are asking her to work harder, or possibly leave her home, or move away from her children and friends. Maybe she's just not interested in what you are interested in and has other ideas and plans. It's not fair to expect her to give up her life and pursue your dreams. How about her dreams?

If people don't really want a life of home canning, gardening, and tending animals, for them, nothing could be more tiresome. The truth is that unless you love it, homesteading or farming can be a mind-numbing life of pure

---

drudgery. It's one of the reasons so many people sell the family farm and move to town. Town life is so much easier.

What I think you are saying is that after years of working for someone else, you want to work for yourself and enjoy life. I agree with you and don't blame you. Why work hard so that you can be paid in a depreciating currency and then pay another person or organization to meet your basic life needs? It seems more sensible to cut out the middlemen and do it yourself.

"It's a hard life - but it's the good life." - Butch Garrett.

The only advice I can offer to you and your wife, or to anyone for that matter, is this:

Go slow and grow into the changes that you want. Nature teaches us that when it comes to living things, gentle, slow, and steady will bring progress. An evolution is better than a revolution. If you already have a situation where you can homestead in town, I'd stay put. As long as you have a small piece of ground and the sun shines on it, you can grow and preserve lots of your food. Chickens, rabbits, and even a dairy goat can provide plenty of fresh milk, meat, and cheese. Manure from the animals can help your garden grow. Don't bother leasing land. Consider investing your money in a water filter for town water and some good garden tools. Explore alternative heat, cooking, lighting, and toilet systems. A freezer, additional pantry space, and household storage will likely be money well spent. Try to connect with other like-minded people. It makes good sense to trade the fruits of your labor for those you aren't able to produce. Now is the time before you actually retire to find ways to cut back on your expenses and build the infrastructure that you'll need in the future. It takes time to acquire the necessary skills and tools to live a more independent life. I would be especially cautious about biting off more than the two of you can chew.

Keep in mind that you are not going to get younger. You are growing older every day, and there will be physical limitations and perhaps illness in the future. Also consider that statistically, unless your wife is 10 to 12 years older than you, she will probably outlive you - so plan ahead.

Don't put her in a future situation where she may have to care for animals, a big garden, food storage, and a sick or dying husband, if it isn't something she really wants for herself.

By staying put and continuing to provide for many of your own basic needs, you may find that your wife discovers for herself the economic benefits and pleasures of the life that you long for. Who knows?

Maybe by this time next year, she'll insist on moving out of town so you can get a couple of pigs and a goat. Anything can happen when love is involved. Good luck to you both in the future.

It's not a secret. I'm no fan of so-called prepping or survivalism. What follows is a question from a few years back regarding prepping and

The Complete Granny Miller

preppers. Some of my objections to the notion of prepping are outlined in my answer to the reader. But what I didn't convey in my answer is my long-held belief that preparedness and prepping are simply the latest installment of historical and traditional American agrarianism. It seems that every 25 years or so, there's some kind of new 'back to the land' or 'simple living' or 'return to my roots' movement. It's a generational thing.

Since the beginning of our nation, American agrarianism has piloted personal and public change. Every rebirth of agrarianism provides the reassurance of traditional values and a hoped-for economic sanctuary for the middle and working classes. Agrarian revivals are almost always a social protest against the status quo. Typically, agrarian movements occur during economic and political change and upheaval. They often contain the seeds of civil disobedience and a mistrust of government. American agrarian revivals predictably follow periods of war or a collective national trauma. During such times, the values and principles of rural life are idealized. In comparison, the perceived insecurity, dangers, and difficulties of city life are rejected. People instinctively flee cities and make a run for the country when times are uncertain. Modern prepping and preppers are the expected and anticipated American agrarian response to the events of September 11, 2001. But most preppers don't know that.

My generation of homesteaders took off during the 1970s. We were the agrarian social response to the War in Vietnam and to other political and social upheavals of the time. Every generation of homesteaders owes a debt to the generation that came before. Homesteaders of the 1970s were influenced by the late John Seymour and by Scott and Helen Nearing. Seymour, the Nearings, and others were part of the agrarian revival that predictably followed World War I and World War II. Their writings heavily influenced the post-World War II generation of American homestead authors and small holders, most notably of these, a young and green Gene Logsdon, and a wet-behind-the-ears Wendell Berry. Both Wendell Berry and Gene Logsdon have rewritten and redefined American agrarianism. Both men inspired and schooled Joel Salatin, Michael Pollan, and many others of my generation.

Back in the 1970s, most homesteaders were 'hippies' who were flocking 'back to the land' to tune out the 'man.' At the time, the 'establishment' was a 'bummer' and the government couldn't be trusted. (*The government still can't be trusted.*) It was during that period of American agrarian regeneration that many intentional communities like The Farm in Tennessee were formed. One difference between this current generation of young preppers and my generation of old hippies is the emphasis upon paramilitary philosophy, consumerism, and technology. My generation's philosophy was the exact opposite. It emphasized a willful ignorance of nature, pacifism, self-entitlement, and disdained materialism.

But philosophy and classifications really don't matter. Because when it comes time to shovel a steaming pile of manure or to pick green beans in the hot sun, labels are positively useless.

The Complete Granny Miller

Prepper Verses Tradition

Dear Granny Miller,

I've recently been exposed to your website for the first time. I've enjoyed what I've seen so far. I actually found a link which led me to you while I was visiting the Rawles survival blog. I'm interested to know why you seem so hostile towards preppers or the prepper movement. Perhaps I'm misinterpreting what I've read. Recently, my great-aunt and uncle (in their 70s) were making fun of a prepper show they'd seen on TV. While some of the 'preps' are doomsdayish and hopefully unnecessary, my kin also made fun of the featured family for having 'stockpiles' of canned goods, water, and weapons. I was surprised by their attitude and stated that it was strange that what was normal only fifty years ago would be seen as outlandish today. That hushed them a bit as they saw my point. So, I agree that a lot of 'prepping' was just normal a while back. I'd like to see it become normal again.

My question to you is, regardless of what it is called, do you agree that being prepared is a good thing? Is your dislike for the movement an issue of semantics or do you really harbor disdain for the newly prepared?

Thanks
James

Dear James,

Many people are astonished to discover that I'm no fan of survivalism or prepping. I'm not opposed to self-dependency and household sustainability. I encourage it. I've respect for people who are cognizant of the realities of this worldly existence and those who desire a life of greater self-awareness and independence. My objections to prepping and survivalism are philosophical and have little to do with the outward material expression of preparedness.

"The limits of my language are the limits of my world." - Ludwig Wittgenstein.

My knee-jerk reaction to so-called survivalism and preparedness is, in truth, partly grounded in the issue of language and semantics. Language not only accurately conveys culture and cultural ideals, but language and the words we employ shape our worldview, cognition, and behavior. Language is constantly evolving. Once the meaning of a word is altered to embrace a new or changing concept, there's no returning to the prior interpretation. Society is permanently and irrevocably transformed, and we along with it. When it comes to prepping or preparedness, I find it instructive that ex-military men run so many popular survivalist websites. In the terminology of those who are actively prepping, my pantry and cupboard now contain 'preps'instead of food and medicine. The extra cheesecloth I keep on hand is an 'alpha strategy' that I'll need when the 'SHTF.' I possess a 'tactical advantage' because I keep a 12-gauge shotgun behind the kitchen door. Friends, neighbors, and the UPS deliveryman are potentially 'zombies' that could threaten me or my family because they may not be 'prepared'. Or

The Complete Granny Miller

worse, they may have witnessed me hanging out laundry, which is a breach of off-grid 'operational security'. My farmhouse is considered to be a 'secure isolated retreat' or an 'emergency BOL' for my children. When English peas and tomatoes are producing in the garden, I don't just have something good to eat for supper, but instead I have 'food insurance.' Commonly used prepper and survivalist terms communicate and express far more testosterone and masculine energy than I care to employ in my homemaking pursuits and endeavors.

"The entire modern deification of survival, per se, survival returning to itself, survival naked and abstract with the denial of any subsequent excellence in what survives except the capacity for more survival still, is surely the strangest intellectual stopping place ever proposed by one man to another." - Colonel Jeff Cooper.

To my way of thinking, the notion of paramilitary household management is the antithesis of stability, security, and comfort that most people desire in a home. I am positively revolted by the survivalist perception and belief that somehow there will not be enough food, water, clothing, shelter, goodness, and humanity to go around when 'the collapse' comes.

Such valuations are indicative of people who are consumers rather than producers. The military is rife with such persons. As far as I'm concerned, it's bad enough that our entire country has become militarized. But now, according to preppers and survivalists, my home has become a potential bunker, and my personal domestic economy a battlefield.

Those who are survivalist-oriented will insist that I'm a Pollyanna because I refuse to acknowledge the dangers of the 'coming collapse'. They will assert that I should trade my apron and garden hat for body armor and infrared night vision goggles. But since when was mere survival the ultimate goal of human life?

Many people who read my opinions will probably disagree with them. But I would tell my critics that TEOTWAWKI (The End of The World as We Know It) happened years ago.

The economic and social collapse is not coming. It is here and it's now. This is what it looks like. Preppers and survivalists have only just recently got the memo.

The Complete Granny Miller

Chapter 2

"God always takes the simplest way." - Albert Einstein

Non-Electric Living

Actually, there is really no such thing as genuine self-reliance or self-sufficiency. Human beings in their natural state are fragile creatures. Left alone and at the mercy of the natural world, ours is a tenuous existence. All of us depend upon the whims of nature, on other people, and on technology in one form or another to exist in this world. Each and every one of us lives by the innovations and inventions of those who came before us.

But not all technology serves to improve human life. There is a tangible value in making informed and attentive choices regarding the technologies we choose to employ to improve our daily lives. This is especially true for homesteaders or for those desiring a socially ethical or prudent way of life. Every technological innovation and achievement requires a period of testing to understand the impact that it will have upon our lives and our communities.

The Old Order Amish here in Pennsylvania are well known for recognizing potential problems that untested or new technologies may bring to their families and communities. American culture, to a great extent, has embraced the idea of an ever-evolving and possible technological utopia. It's a seductive idea. That technology can somehow perfect our human nature and remedy the problems of this world. Inherent in the concept of an ever-evolving material progression is the suggestion that physical or mental labor is somehow something to be avoided at all costs. Or that older technologies and innovations need to be replaced simply because they are long-standing.

The True Cost of Heating with Wood & Coal

I live in rural Western Pennsylvania, where I heat and cook with wood and coal for about 8 months of the year. For the most part, I don't burn much coal because wood works well for me. Only during the hardest and coldest part of winter do I use a bit of bagged anthracite coal for heat. Solar heat or energy is not really possible or realistic for my home. There are too many cloudy days here in Western Pennsylvania during the winter months. When given a choice, I've always preferred affordable, low-tech options that I can easily understand.

IT'S A TRADE-OFF

Where I live, hardwood is plentiful, and few state regulations govern the operation of solid fuel appliances. There are, however, federal regulations

The Complete Granny Miller

regarding wood or coal-burning appliances. But they are often ignored at the local level whenever possible. Wood and coal have been an affordable energy alternative for my home when compared to natural gas, electricity, or petroleum.

Over the years, I have literally saved tens of thousands of dollars by heating and cooking with wood. My wood and coal-burning stoves have paid for themselves at least four or five times over. And that includes the cost of the expensive cook stove in the kitchen. This year (2010), I will probably save $3600 to $3800 in heating costs. All contingent upon how cold and long this coming winter is.

The money saved this winter in heating costs is the price of a middle to top-of-the-line wood stove or furnace. The energy savings are more than half the price of the most expensive cook stove I'm aware of. So, depending upon the brand and model chosen, a solid fuel stove at today's prices will pay for itself in saved energy costs within the first year. The money I'll save doesn't include the cost of propane for the LP gas stove in my kitchen that won't be used again regularly until next summer. Back in the days when I had an electric range, the energy costs averaged about a quarter of my total electric bill every month.

Here in rural Western Pennsylvania, free wood can often be had if families are willing to spend about a month of long, hard Saturdays cleaning up slash wood from commercial logging operations. Gathering free firewood is a good exercise. It always seemed to me a better use of family time and resources than going to the gym or hauling children to extracurricular activities. All it takes is a few phone calls, the cost of a chainsaw, and a willingness to work hard. For those who cannot cut their wood, seasoned firewood at present in my area of the country is running about $150 (2010) a cord delivered. Firewood is measured in 'cord wood'. A cord of wood is a stack of wood 4 ft. deep by 4 ft. high by 8 ft. long. It takes 5 to 7 cords of seasoned firewood for my house to make it through a winter.

My husband cuts and splits most all the wood for our home. It's a big job for one man and usually takes him about three or four complete weekends working 8 hours per day. Time can be saved if the trees are already on the ground. But if trees need to be dropped, it can take much longer. Felling trees and removing the branches takes time and planning. It can be dangerous, hot, and dirty work.

The freedom from dependence on big energy companies and the need for good weather to enjoy a hot meal and a warm home is a source of comfort and security. The fact is, nothing keeps you as warm as wood or coal heat. But heating and cooking with wood or coal come with other costs that are often unseen and unknown to the general public. In my life, there have been real tradeoffs in terms of time, labor, convenience, and lifestyle. No matter how you look at it, there's no free lunch. If you live in a cold climate and want to avoid freezing, you'll need to earn money to afford energy-efficient heating. Or you have to be willing to adjust to a lifestyle of labor and discipline that many people find confining and sheer drudgery.

WOOD or COAL

I have three solid fuel stoves in my house. Two of them have small fireboxes. So that means that during hard winter, I can't be gone from home for more than five hours or so unless I'm burning coal. That's because the wood fires will begin to go out if left unattended. In the early 19th century, when wood was the only option for most of rural America, someone usually had to stay behind at home to keep the home fires burning. Nowadays, to restart a fire is not a hardship because of matches and newspapers.

Before the invention of matches, to have a fire go out was nothing less than a small household crisis. A dead fire usually necessitated having to strike a spark from flint and steel and hope for good luck. Often, a child was sent to the neighbor's house to bring live embers home, with sometimes disastrous consequences. Many a child was seriously burned due to immaturity or carelessness while carrying hot embers. Back then, to have a fire go out meant waiting in the darkness and cold until the fire could be started again. Without a fire, there was no cooking or hot water for cleaning or personal hygiene. It could take an entire day to remedy the situation and to get the household back to running smoothly again.

When burning coal, I can be gone from home for a much longer amount of time. But anthracite coal is expensive. At present, bagged anthracite is $6 a bag or around $300 per ton (2010). If I were to choose only to burn anthracite coal for heat, I would use no more than 1 ½ tons of coal a year. So, this year, my heating costs would be around $450. If I were to select bituminous coal instead of anthracite, my heating costs would be even lower. Bituminous coal is a lower grade of coal. It is dusty, smelly, and does not burn as clean or as hot as anthracite coal. But it is readily available here in Western Pennsylvania. If I supplement my wood burning with coal, I will typically use between 1/2 to 1 full bag of coal per day, depending upon how cold it is. The closer the thermometer gets to 0°F, the more coal I would have to use. But even at the price of $6 a bag or $300 a ton, coal is a considerable savings over fuel oil, electricity, or natural gas. If I lived in an area where wood was not readily available or if I worked away from home, I would probably choose coal as my primary heat source. Due to the small fireboxes in my stoves, I have had to cut many winter trips short to return home and feed the fire. Those moments are inconvenient. But I'd not be telling the whole truth if I didn't also mention that there have been times when I have used the excuse of tending to a wood fire to leave a tedious social situation early.

WHAT A DAMN DIRTY MESS

Heating and cooking with wood have many benefits. But one thing I have never been able to get used to, and I consider a disadvantage, is the sheer amount of dirt, snow, mud, bark, wood chips, insects, and debris associated with wood burning that ends up in the house. Coal is not so bad. But it takes a good started wood fire to make a coal fire. There's no getting around it. Wood has had a direct impact on my housekeeping and has authored the interior design of my home. Wood and coal heat are dirty alternatives to electricity, fuel oil, or natural gas. So, in terms of housekeeping, if I want to stay happy, the best I can do is manage the wood mess in my home, and

The Complete Granny Miller

realistically accommodate the life I choose to live. Years ago, I declined to follow the style of many middle-class American homes just to stay sane during the wood heating months. For my home, that meant scrubbable painted walls and trim; no wall-to-wall carpeting or drapes; washable upholstery; and a big red fire extinguisher in almost every room of my house. Soot transfers easily and can be hard to remove. Soot smudges are a constant battle and can end up on woodwork, the bathroom sink, the refrigerator, and elsewhere.

THE HOUSE CAN GET TOO HOT OR TOO COLD

Heating with wood or coal is not as convenient as simply flipping a switch or turning up the wall thermostat. Heat from a solid fuel appliance is much more creaturely comfortable, but it's not a stable heat. A modern natural gas or fuel oil furnace, or an electric or hot water baseboard heat system, is more consistent. Wood heat always needs to be fiddled with and is a drying heat. No matter how many pans of water I set out, the relative humidity in my house rarely rises above 30% during hard winter. That kind of desert-like dryness takes a toll on wooden furniture, books, and skin. Most days when I'm busy around the house, I work in a tank top because the house seems too hot to me, with temperatures averaging 80°F - 85°F. But on winter mornings when outdoor temperatures are in the single digits or lower, and the house has lost temperature overnight, it is sometimes so cold in my bedroom that I can see my breath. Many mornings, I've lain in bed under the warm down comforter, hoping perhaps some compassionate wood stove fairy would take pity and get the fires going again. To get out of bed on a bitter cold Western Pennsylvania morning takes no small amount of courage.

I WOULDN'T HAVE IT ANY OTHER WAY

Usually by the end of February, I've had enough of wood and coal and all the things that go with them. March and April are challenging months to heat with wood because of the approaching spring. Often, the weather is cold only in the morning or evening, so fires don't need to burn all day. That can lead to starting two or sometimes three fires a day in the same stove. When the weather is cold and rainy during the spring, it can be difficult to know how long, how hot, or how many stoves need to be fired up.

When I first started heating and then later cooking with wood, it took me about three full years to understand all the variables in stove operation. When heating with wood or coal, nothing can replace personal experience. You must live it to understand it. In the beginning, I had to learn a new way of living and adjust my attitude and outlook to a new cycle of life. A life centered on and around the tending of a fire. The notion of hearth and home took on deeper meaning for me. Many years have gone by since then. I'm old now and well-seasoned - just like good firewood. I have been heating with wood for so long that I can hardly remember any other way of life. In spite of the labor and mess involved and the sometimes terribly cold mornings, I wouldn't have it any other way. Heating with wood and coal is right for me and my circumstances. It has made good economic sense over the years and brings with it a measure of independence and energy security

The Complete Granny Miller

that wasn't found back when I was dependent upon the energy grid. But believe me, all that said, the spring is always welcomed around here.

Heating Your Home With Coal

Don't believe everything you hear or read about coal. Coal has gotten a pretty bad rap in the last 25 years or so. Big petroleum energy corporations hate coal, as do the true believers of the sham science and religion popularly known as 'Climate Change'. The truth is, modern coal-fired appliances are efficient and affordable.For people who live in town, coal heat can be a wise choice. Unlike wood, coal does not attract insects and can be stored indefinitely. During the coldest part of winter, I often burn coal instead of wood in my stoves.

There are two main types of coal used in the United States for energy:

- Bituminous coal
- Anthracite coal.

Bituminous coal is the most common form of coal. It is the type of coal used for electric power plants, but it is also used for home heating. Bituminous coal is dull and dusty-looking and is easily burned. It is considered to be a soft coal and burns sooty. Bituminous coal contains 10,500 to 14,500 BTUs per pound.

Anthracite coal is a denser, harder coal that is found in the US, but only in Eastern Pennsylvania. Anthracite coal is about twice as expensive as bituminous coal. It's almost always used for home heating and not for electricity generation. Anthracite coal is shiny and waxy-looking, and it burns remarkably clean. Anthracite coal contains about 15,000 BTUs per pound.

The advantages of coal for home heating are many. Coal can be safely stored for an indefinite period, and it never goes bad. Coal doesn't rot or draw insects like wood. Coal does not require a pipeline or any special tank or container, unlike LP gas or fuel oil. Depending upon your location, coal is often a more affordable home heating option when compared to either fuel oil or natural gas.

Best of all, coal is not produced by people who want to behead you or hate you. Coal is a 100% American energy resource. Coal is abundant in the United States, and many coal mines are still small Mom and Pop operations. Coal makes jobs for Americans. The last I knew, Pennsylvania has enough anthracite coal for home heating needs for at least another 150 years or so - maybe longer. A coal fire puts out more BTUs than most types of hardwoods. Osage wood is the only wood I can think of that will burn as hot as coal. Coal is readily available in many areas of the country and is most

---

The Complete Granny Miller

often sold in 40lb bags or in bulk.

Now, before you rush out and buy coal for your wood-burning stove, there are a few things you need to know. To safely and effectively burn coal, you must have a multi-fuel stove or appliance fabricated specifically for coal. Coal fires burn too hot for most standard wood stove fuel boxes. Burning coal in a regular wood stove or furnace can result in an overheated or overfired stove. That means a burned-out fuel box, a warped stove, or a house fire.

All coal stoves or furnaces have a way to bring air to the fire from underneath. Coal appliances are designed with a cast-iron slotted grate. They all have a way to shake or tip the grate, clearing out ashes and any leftover coal clinkers. A coal fire must have free circulating air from beneath in order to burn properly. Any buildup of ashes under the grate will inhibit a coal fire.

HOW TO START AND MAINTAIN A COAL FIRE

Coal fires, unlike wood fires, can be hard to start and need to be tended to differently. I'm going to assume that if you're interested in burning coal that you already know how to start a fire in a stove or a furnace. I'm also going to assume that you have a multi-fuel stove or appliance. Along with a poker, you'll probably want a coal hod and a small shovel to manage an indoor coal fire. A coal hod is also called a coal scuttle or coal bucket. Coal hods frequently have a pitcher-shaped end for pouring coal on a fire. Coal hods are usually made of metal and have a handle for carrying small amounts of coal.

To start a coal fire, you'll first need to have a good, strong wood fire going. Depending upon the type of coal you plan to burn, you'll need a bed of hot wood coals. For bituminous coal, it's about a 1-inch to 2-inch bed of wood coal. With anthracite coal, a bed of hot wood coals about 2 to 4 inches deep is usually required to get the coal fire started.

With both types of coal, the coal fire is started by adding a small amount of coal on top of the wood coals. The lower door or damper of the stove must be opened so that coal can be fed air from beneath. After placing the initial small amount of coal on top of the wood coals, wait about 5 minutes, then add twice as much coal as before. After about 10 to 15 minutes add more coal and begin to watch for the blue flame that is characteristic of coal-burning. Once you see the blue flame, you can close the bottom door or damper. When there is a full bed of ignited coals on the grate, an entire hod of coal can be added. Remember, add the coal by the hod while the damper or bottom door is open. Pour the entire hod of coal onto the fire. Wait to make sure that blue flames are creeping up through the coal before closing the bottom door or damper.

Many people who become frustrated with coal burning fail to appreciate the differences between the two types of coal. With anthracite coal, it's essential to neither rush the coal ignition nor stir up or poke the fire like a wood fire. An anthracite fire needs to have the grate gently shaken or lifted slightly and moved every once in a while. If you disturb a mass of anthracite coal by poking or stirring it, the fire will tend to go out, and you'll be left with unfired clinkers.

The Complete Granny Miller

With a bituminous coal fire, the coal will tend to burn and lump together into a large solid mass. Bituminous coal fires need to be lightly poked and stirred up in order to burn completely. The most important thing to understand about burning coal is that it doesn't burn like wood. Coal radiates and burns from the bottom to the top. The fire spreads upwards through the coal. One piece of coal will ignite another. When a coal fire is properly burning, there is little flame. The coals just glow red. An entire hod of anthracite coal will keep my 1200 square ft. house comfortably warm in -15F° weather for about 6 hours. In sub-zero weather, I will use just over one bag of anthracite coal a day. I never burn bituminous coal in the living area of my house because of the soot and odor. However, I do burn bituminous coal in my cellar stove to keep the plumbing from freezing in sub-zero weather. Anthracite coal has a more complete combustion than bituminous coal. Anthracite coal leaves little ash and waste when compared to bituminous coal.

Unlike hardwood ashes, I don't spread coal ashes on my garden. Instead, coal ashes are scattered in my driveway to melt ice and snow. Some people prefer to start coal fires with outdoor charcoal briquets. I don't recommend the briquet method due to the costs of the briquets and the lack of availability of charcoal briquets in some areas of the U.S. during the winter months.

Hand Protection When Loading Wood Stoves

There's a pair of heavy welder's gloves next to every wood and coal stove in my house. I never load a wood stove or open a stove door without wearing them. I learned about hand protection and wood stoves the hard way 10 years ago when I suffered a 2-inch-long third-degree burn to my wrist while loading a stove. Gloves are cheap insurance.

Steel Ash Carrier - Keep Wood Ashes Safe

Believe it or not, seemingly cold ashes, cinders, embers, and stove coals can sometimes keep their heat and reignite for up to a week or longer once a fire goes out. Many an accidental house fire has been the result of the improper disposal of ash or coal clinkers. The tragic 2011 Stamford, Connecticut Christmas fire in which a woman lost her three children and parents is a grim example of how ashes can seem to be dead but aren't.

In my own life, I've had a couple of close calls with ash and cinders sitting forgotten in a coal hod in the living room. There's an indescribable horror to walking into a room filled with papers, books, and cloth, and seeing a coal hod of ashes glowing radiant and cherry red hot next to a wooden bookcase. A steel ash carrier will keep coals and ashes contained and safe. They can be lifesavers. An ash carrier with feet will keep the heat of the ash and live coals off the floor and carpet. So, there's less chance of burning or scorching a wooden floor or rug.

The Complete Granny Miller

Some stove manufactures sell ash carriers as a stove accessory. Prices range from about $45 to $75 (2010), the last time I checked. An ash carrier that fits the stove ash pan will help prevent fly ash from blowing into a room when cleaning ash from the stove.

Hardwood ashes are a valued commodity on my farm. Ashes can be used to make soap, clean wooden floors, shine silver, melt ice, and are helpful in the garden. Without a doubt, the garden is my favorite place to use wood ashes. They are most often used to top dress rows of asparagus. Hardwood ash is rich in potassium, calcium, and other minerals. Asparagus thrives when hardwood ashes are applied to the soil. Tomatoes love hardwood ash when they are put directly into the hole at planting time. But not all plants will benefit. Hardwood ash should never be applied to a potato patch, as it may contribute to the development of potato scab.

Cook Stove Basics

I use a wood-fired cook stove for about 8 months out of the year. A cook stove is a lot of work. But from my point of view, it's also a lot of independence and security, too. No matter what the weather brings or what happens with energy prices, I will always be able to heat my home and cook for practically nothing, just as long as I am willing to pick up sticks and split firewood. And when I grow too old to split wood, I can always burn coal if I choose.

My cook stove is a traditional Waterford Stanley. It's a modern solid fuel stove. A Stanley will burn peat or wood, and with a change of firebox liner, it will burn coal. My stove not only cooks but also helps to heat my home. The Stanley has a place to plumb a pipe for hot water, should I want that. All cook stoves have individual differences, but basically work the same way.

In general, wood cook stoves are similar to other wood-fired appliances. They are connected to a chimney and have a firebox. They all have some way to clean out the ash, and air to the fire is controlled by some type of baffle(s) or damper(s) system. Cook stoves, unlike regular wood stoves, have an oven. The oven in a cook stove is simply a box within a box. Some cook stoves have a water reservoir attached to the side for hot water. A water reservoir is handy to have, but it requires more maintenance than a cook stove without one. The water reservoir must never be allowed to run dry because it will ruin the silver solder seams and cause the reservoir to leak.

Many cook stoves have a top compartment called a warmer. The warmer is used to keep food or plates warm and for dehydrating food. I use mine to defrost food and warm up my hat and mittens during the winter. The temperature of the warmer depends upon the temperature of the stove pipe

The Complete Granny Miller

and the top of the stove. Most of the time, my warmer stays a cozy 130 F. Back before the days of electric or modern solar food dehydrators, the stove warmer was a good place during cool, rainy fall weather to dry apples, pumpkin rings, and other garden produce for winter storage. Using a warmer was a significant improvement over previous methods, which consisted of drying food in the sun. Or in the case of apples and pumpkins, slicing them into ring shapes, and then stringing them across the kitchen or in an attic to dry.

The top surface of a cook stove is called the hob. There are round plates on the hob called 'eyes' that are removable for cleaning out soot and ash. They are not burners. With some cook stoves (not mine), the eyes can be removed while the stove is in operation. Stove eyes are removed to add small pieces of wood directly to the fire. Or to set a pot into the hole for more heat if you need it to cook faster. The eyes are lifted off with a 'lifter'. On a cook stove, the entire hob is used for cooking. The part of the hob that is directly over the firebox is the hottest part of the stove. When cooking on most cook stoves, the pots and pans are moved from left to right to control the cooking temperature. The left side of the stove is the hottest part. The right side of the stove is the coolest.

The temperature of a wood cook stove is regulated in a few different ways. The temperature is controlled by the type and amount of wood the fire is fed. It's also regulated by the amount of air that the fire receives and by the size of the fire. The type of wood species used in a cook stove significantly affects the fire's heat and duration. Small, dry pieces of wood are best for fast fires. Poplar or pine burns cool and is considered 'summer wood'. It was used quite a bit during canning season years ago in summer kitchens. It burns quick and will leave a lot of ash. In my stove, I use maple and sometimes cherry for my everyday cooking, because that's what's plentiful on my farm. Maple will build heat quickly, but it won't last long. Hickory or oak burns especially hot and throws off a lot of heat. Both woods make a good fire for frying or rapidly boiling water. Oak and hickory tend to overheat the stove and will make the oven too hot for most baking or roasting. Some cooks avoid oak and hickory altogether because they tend to burn out a firebox. My favorite wood for rapid heat is apple wood.

A cook stove's temperature can also be controlled by how much air the fire receives. For a quick-burning fire, open the bottom or firebox damper(s) to allow more air to the fire. For a long-burning fire, close the damper(s). To hold a fire from one meal to the next, simply put a large piece of wood into the firebox after cooking is finished and close the bottom damper(s). The wood will burn, but it will burn slowly. When it's time to cook again, open the firebox damper(s) to give the fire more air and add wood to start the fire burning hot again. To let off heat from the firebox and cool the stove down, open the chimney damper and allow the heat to go up the chimney.

For most cook stoves it takes about 15 to 20 minutes for the hob to heat up and be ready for cooking when first started. For baking, the stove usually

The Complete Granny Miller

needs at least 45 minutes to 1 hour before the oven is ready. As a general rule, if the top of the stove is hot enough to boil water, the oven will be hot enough to bake.

Oven temperature is controlled by a separate damper(s) that work to hold the heat in the area that surrounds the oven. To use the oven, bring the oven temperature to within 25° to 40° of the desired setting. Then close the oven damper(s). After about 10 to 15 minutes with the oven damper(s) closed, heat will have built up inside the oven, and the oven temperature will hold steady. To quickly reduce the oven temperature, simply open the oven door and let the heat escape. Once the oven temperature is where it should be, it can be maintained by adding just a couple of small pieces of wood whenever the temperature starts to fluctuate.

It's essential to keep in mind that a cook stove oven doesn't heat the same way a modern electric or gas range does. Heat collects at the top of a cook stove oven. This is good for breads - but a disaster for cakes. A trick to prevent a cake from burning is to place it on the bottom or floor of the oven, with a pan of water on the top shelf. The pan of water collects the heat and acts as a barrier to the top of the cake. Also, the side of the oven closest to the firebox will always be hotter. Food in the oven needs to be turned often to prevent burning on one side.

A well-made cook stove will last more than a lifetime. Cook stoves are safe if they are well cared for and common sense is used. A cook stove must be cleaned and inspected regularly. Built-up ash under the hob and around the oven will reduce the temperature of the hob surface and the oven. Ashes and soot should be cleaned out from underneath the oven and the hob frequently. Any ash along the interior sides of the stove walls should be swept out. Generally, cook stoves are cleaned from the top of the hob, then along the interior sides, and then ash is raked out the bottom at the soot door. A soot rake is used to clean the back of the stove and access all hard-to-reach areas. I clean out the interior ash and soot from my cook stove about every 3 weeks or so. Some people will clean more often. The chimney thimble and stove pipe that service a cook stove should be cleaned and swept out about 2 or 3 times during the wood-burning season. Occasionally knocking gently along the length of the stove pipe when it is cool will help to knock off any collected creosote or ash that may have built up between cleanings. Debris that falls off the stove pipe falls downward. It will collect under the hob where it can be swept out during routine cleaning. If chimney fires are to be avoided, it's crucial to keep all parts of a wood-burning appliance clean. I think cleaning a cook stove is probably the worst part of owning one. It is a dirty job. A heavy-duty brush, newspapers, rags, rubber gloves, and a bucket of ammonia water are usually necessary.

And speaking of chimneys and cook stoves, a chimney that is subject to a downdraft can make some days miserable. My kitchen chimney is subject to downdrafts. Since my cook stove isn't airtight, it tends to smoke and back puff on windy days, especially when the wind blows strongly from the southwest. A Vacu-Stack chimney cap helps some. But on the windiest of days, the best I can do is open a window.

Just so you know, not all modern cook stoves have an old-fashioned or

The Complete Granny Miller

nostalgia design. There are quite a few modern and beautiful designed contemporary cook stoves. I chose my stove because the hob is 34" high, which is comfortable for me to work at, and because the Waterford Stanley only requires a 4-inch wall clearance with a double-insulated stove pipe. I also chose my stove because it's solid cast iron and not sheet metal. It retains and radiates heat for a long time.

If you are considering buying a used cook stove, keep in mind that antiques look great but may have cracks or other problems. Many good cook stoves from the 1940s and 1950s are still out there. Some are in pretty good condition and are reasonably priced. Cook stoves require some effort and trouble. But for what you get back, I think they're more than worth it.

Cook Stove Warmer

"Under this is placed a closet for warming and keeping hot the dishes, vegetables,meats, etc., while preparing for dinner. It is also very useful in drying fruit."

THE AMERICAN WOMAN'S HOME –
1869 Catherine E. Beecher & Harriet Beecher Stowe

Many modern and antique wood-burning cook stoves have a top compartment known as a warmer, a bun warmer, a plate warmer, or a warming closet. My cook stove is no exception. The purpose of the stove warmer is precisely what the names imply. It's a place to keep things warm. A warmer will, of course, keep rolls, biscuits, and bread warm. In some ways, it's a glorified crisper. I usually keep a set of serving bowls and plates in the warmer so that they'll be warm when I need one for supper. Warm plates and bowls keep food hot longer. Nothing is worse than cool mashed potatoes.

A cook stove warmer will also keep a plate loaded with food hot for a few hours, just waiting and ready for whoever missed supper. To keep a warm supper waiting during a snowstorm is an act of faith that a loved one will eventually return home safely. I often use the cook stove warmer for defrosting food, cooking rice, and keeping a pot of tea warm. But my favorite use for the warmer has nothing to do with food. It's the place where I dry my hat and mittens during lambing season. Dry warm mittens make it easier to go back outside and face winter weather barn chores.

The temperature of the warmer depends upon the temperature of the stove pipe and the top of the stove. Most of the time, my warmer is a cozy 130° F. But if I'm burning the stove hot for an extended period, the warmer can get too hot for me to touch the inside comfortably.

Back before the days of electric or modern solar food dehydrators, the stove

---

The Complete Granny Miller

warmer was a good place during rainy fall weather to dry apples, pumpkin rings, and other garden produce for winter storage. Using a warmer was a significant improvement over previously methods. Those methods consisted of drying food in the sun, or strings of pumpkin and apples across the kitchen or in an attic to dry. Humid or rainy weather could mean mold and spoilage. The stove warmer was a dependable way to dry food, especially when Mother Nature didn't want to cooperate.

Apples were prepared by peeling them first and then slicing them about 1inch thick. Next, they were soaked in a mildly acidic solution to prevent them from turning brown, usually in cider vinegar. But lemon juice was used in some cases. The apples were then arranged upon a tin or plate and set in the warmer. The apples were turned every 3 or 4 hours. Often the apples were left out to dry overnight. Slices of pumpkin and other fruits were preserved in the same way. Dried beef, corn, and green beans were also favorites to dehydrate in a stove warmer.

Preventing Rust on a Wood-Fired Cook Stove

The best way I've found to keep rust away from a cook stove hob is with a light layer of vegetable shortening or cooking oil. Many cook stoves are prone to drawing rust all over. And not just on the hob or the top surface. Rust is more of a problem during the summer months than during the winter months. During the summer, it is humid, and the cook stove sits idle. During the winter months, the cook stove is in use every day and doesn't have a chance to get many rust spots, but it will occasionally get a few. The rust spots are mostly rings from where a wet pan was set down when the stove was cool.

To keep rust at bay, apply a bit of vegetable shortening or cooking oil to the top surface area and the hob of the cook stove every few weeks or so using a clean, lint-free cloth or wax paper. If the stove is cast iron, apply shortening along the sides and doors, too. Apply anywhere where rust may be a problem.

It feels like polishing furniture. But in reality, it's more like oiling a gun. It takes a while for the grease to burn off from the hob. It's easy to tell when I have polished the stove. Not only does the cook stove look better. But for two days afterwards, the whole house smells like I'm making pancakes as the shortening slowly burns off.

Chamber Pots

Like many older farmhouses, my home was constructed without indoor plumbing. But in the early 1970s, indoor plumbing was added to my home. That's pretty late by American standards. Up until that time, all water for drinking, cooking, and bathing was hand-carried into the house. Water was also carried in by buckets to the cellar for laundry, which was heated by fire in a copper boiler.

The Complete Granny Miller

For most of the family's toilet needs an outhouse was located in the backyard. The outdoor privy has since been torn down, but a foundation stone remains to mark the spot. For nighttime toilet needs, a chamber pot was used.

A chamber pot is also known as a piss pot, a jerry, a jordan, a thunder pot, and probably by several other names I'm not familiar with. Up until the introduction of indoor plumbing, most people would keep a chamber pot in their bedroom for nighttime convenience. The pot was kept under the bed, in a nightstand, a washstand, or a basin stand, and was emptied in the morning.

I have early childhood memories of hearing my grandmother using a chamber pot at night. I don't know why she used a chamber pot instead of the toilet when the bathroom was right next door to her bedroom. Maybe she was afraid of waking up the house? Maybe old habits die hard, or she preferred it? I really don't know.

But what I do know is that, like my grandmother, once I reached my mid-50s, I, too, needed to answer the call of nature at least once at night. The problem was that I was sleeping on the second floor, which had no toilet. So, for the next 10 years, I chose to use a chamber pot until I could afford to have a toilet installed on the second floor of my home.

If you find yourself persuaded that a chamber pot may be a temporary or even permanent solution in your home, I offer a few suggestions.

- Old chamber pots can be found at auctions, yard sales, eBay, and elsewhere. Lehman's Hardware used to sell chamber pots, but no longer does. I know there's a need for them. Perhaps American Manufacturers will rise to the occasion and start producing them again.
- If you are going to use a chamber pot, make sure it has a lid. No sense smelling urine all night or splashing urine when you move the pot.
- When the chamber pot has been used, set it out of the way so you don't kick it over when you get out of bed in the morning.
- Ironstone chamber pots never rust and stay fresher than enamel pots, but they can be broken.
- Enamel chamber pails are usually more budget-friendly and have a handle that makes them easy to carry. But they are prone to chipping and rusting.

Advice about Oil Lamps

Hard to believe that less than 100 years ago, most people living in rural America used only oil lamps for their lighting needs. No matter who you are or where you live, keeping an oil lamp or two in your home is a good idea.

Not all oil lamps give the same amount of light or operate in the same way. Here are the fundamentals of what you should know about oil lamps.

Your personal family needs and individual economic considerations should influence the type of lamp that is best for your situation.

As far as oil lamps go, there are basically 3 or 4 different kinds:

- Floating Wick Lamps
- Mantle Lamps
- Flat or Round Wick Lamps
- Pressurized Lamps

Floating Wick Lamps

Floating wick lamps are really just for decorative lighting and emergencies. The light is faint and soft. For the most part, they are safe. Floating wick lamps are based on a design that has been in use for well over 6,000 years. The way that they work is that a piece of cork, bent metal, or other material is fitted with a small wick. The whole rig floats or sits on top of a layer of oil and water. Some people will just use oil in the lamp without the water. The advantage of using oil and water is that should the lamp accidentally overturn, the water will extinguish the flame. Olive oil works best in this type of lamp, and corn oil is almost worse than useless.

Floating wick lamps are similar in principle to early American Betty Lamps. Betty lamps burn animal fat, grease, or oil, and use a simple cloth wick without a floating cork or water. Betty lamps can be challenging to light, especially when the room temperature is below 45°F. In the past, Betty lamps were usually made of wrought iron or ceramic. This allowed the bowl or container to be heated from underneath, melting the grease and helping the lamp light and burn more easily.

Mantle Lamps

Aladdin Lamps are perhaps the best-known mantle lamps. In my opinion, they are the most effective type of oil lamp for general everyday household non-electric lighting needs. You can easily read and work by them without eye strain. A properly lit Aladdin lamp produces the light equivalence of about a 25 to 40-watt electric light bulb. But they are expensive.

Aladdin lamp light is harsh and has a distinctive blue cast to it. They make a faint humming sound when in operation. The way that a mantle lamp works is by the combustion of volatile gases moving across the knitted webbed mantle via a round, continuous tube-shaped wick and flame spreader. A mantle is a round and sometimes conical-shaped knitted cloth mesh that is attached to the central burner of a lamp. The cloth mesh of the mantle is flimsy and is saturated with radioactive thorium or other rare earth salts. The first time a mantle is lit, the radioactive components are burned away, making the mantle extremely fragile but incandescent upon exposure to heat. This is why mantle lamps provide so much light.

Mantle lamps are safe. But as with all open-flame lighting, common sense

---

and caution must be used. The top 18 to 24-inch area around the chimney of an Aladdin lamp gets hot. The chimney stays hot for a long time even after the lamp has been extinguished. In fact, the entire gallery assembly of an Aladdin lamp gets extremely hot. So be careful!

Like all mantle lamps, the flame of an Aladdin lamp will tend to creep higher if it is turned up too high and too fast. When the lamp is turned up too fast, it can cause sooting and black spots on the mantle. But sooting is an easy fix. Simply allow the lamp to cool, relight it, and let the soot burn off. An over-fired mantle lamp can be dangerous and can lead to a 'runaway lamp'. A runaway lamp is a lamp that burns uncontrollably. The best way to deal with a runaway lamp is to turn down the wick and place an empty tin can over the chimney. The can will starve the fire of air.

Aladdin lamps require close supervision when used around children or individuals unfamiliar with their operation. Most Aladdin lamps benefit from a shade. A shade moderates the bright light and directs it downwards towards a work or reading area. Glass shades have the advantage of being easily washable. The downside is that they are expensive and can be broken. Cloth or parchment shades are an affordable alternative. They are not as heavy as glass, and they can be easily covered with any fabric.

Flat Wick or Round Wick Lamps

These are the type of oil lamps that most people are familiar with. The light is soft, quiet, and soothing. The way that a flat wick lamp works is similar to a floating wick lamp. The difference is that the wick is much larger and stationary, and is threaded through a brass or nickel burner.

The flat wick burner is fitted with little teeth or gears that allow the wick to be turned up by a round knob. The lamp fuel is drawn up through the cloth wick by capillary action and is burned off. The higher the wick is turned up, the higher the flame. Wick height determines the amount of light. One problem with a flat wick lamp is that the wick can only be turned up just so far before the lamp starts to smoke, and the flame possibly breaks the chimney. A flat wick lamp has the lighting equivalency of a small electric nightlight. All flat wick lamps benefit from having their wicks occasionally trimmed of carbon deposits and cleaned. Many people use ordinary kerosene for fuel in their flat wick lamps without any problem. But kerosene can give some people a headache. Ultra-Pure Liquid Paraffin, K-1 Kerosene, and Aladdin Lamp Oil are better choices for sensitive people. Those fuels will burn cleaner and produce less odor. However, sometimes

The Complete Granny Miller

there is a noticeable odor after the lamp is blown out.

Round wick lamps appear to provide slightly more light than flat wick lamps. They can be turned up higher without sooting and smoking. Flat-wick or round-wick lamps are easy to use, but don't give enough light to read by. And just so you know, there is a type of lamp called a double-wick lamp. It works just like a single wick, except there are two wicks attached to the burner. In theory, a double-wick lamp gives off twice the light.

Pressurized Lamps

I have limited experience with pressurized lamps. They are popular with the local Amish here in Western Pennsylvania. Petromax, Coleman, and BriteLyt are the two brands I'm familiar with. Like Aladdin lamps, pressurized lanterns are expensive to buy, but they are cost-effective to run, safe, and very dependable. Be aware that there is a learning curve. Unlike Aladdin lamps, pressurized lanterns must be used with adequate ventilation. Pressurized lamps use a gas generator and a gas mantle. They have to be pumped by hand to create the interior pressure and can be a little tricky to operate. Some people find the hissing noise that they make disagreeable. But some people find it soothing. The light is very bright and harsh.

Treadle Sewing Machine Advice

You love to sew. Or perhaps you are looking for a sensible off-grid sewing machine and think you'd like to buy a treadle sewing machine, but don't know where to start or what to look for? Maybe you're worried all treadle sewing machines are expensive antiques and you can't afford one? Or you're concerned that you'll have to do without a zigzag stitch or machine-made buttonholes if you use one? Or maybe you don't know how to sew but would like to learn?

Well, grab a spool of thread and get ready to sew! Because I'm about to give you some practical and basic advice on one of my favorite off-grid topics: treadle sewing machines!

WHAT IS A TREADLE SEWING MACHINE?

A treadle sewing machine is simply a sewing machine that is powered by what you ate for breakfast instead of electricity. All sewing machines have two main elements in common. Sewing machines, whether they are electric or treadle, consist of a machine head and some type of mechanism that drives the head. The machine head is the part of the sewing machine that

The Complete Granny Miller

sews. A sewing machine head consists of precisely machined and tooled fitted rods, screws, wheels, springs, disks, gears, and other parts. Some of those parts are hidden and encased within the head, while some parts are visible on the outside of the sewing head. Keep this information about sewing machine heads in mind because you'll need it later.

The mechanism that drives a sewing machine head can either be electric or non-electric, as in a treadle or hand-cranked sewing machine. An electric sewing machine usually has a machine head with an attached light. The sewing machine may or may not be computerized and is driven by an electric motor.

Treadle sewing machines have two main elements: the sewing machine head and the treadle base. The treadle base is the table or cabinet that the sewing machine sits in. Treadle sewing machines are powered by a drive belt that is most often made of leather and connected to a treadle assembly. The belt sits in a groove on the hand or balance wheel of the sewing machine head. It is fitted down through the top of the sewing machine's table or cabinet base in a continuous loop.

The leather drive band loop usually encircles a large metal grooved wheel under the base of the sewing machine that is attached by way of a pitman rod to a foot treadle. When the foot treadle is worked, the attached pitman rod turns the large grooved assembly wheel. This action begins to move the leather drive belt caught in the sewing machine's hand wheel, and the parts of the sewing machine head begin to move. The result is that if the sewing machine head has a needle and the head is threaded correctly, when fabric is placed under the foot of the sewing machine, sewing commences.

A hand-cranked sewing machine is also a people-powered sewing machine. Instead of being belt-driven, it has a handle attached to the balance wheel. As you turn the handle on the hand wheel, the machine sews. Hand-cranked sewing machines are a good choice for people who don't sew often. They can be quite a bit slower to sew with as opposed to a regular treadle sewing machine. Hand-cranked sewing machines are usually about ¾ the size of a standard sewing machine.

WHO USES A TREADLE SEWING MACHINE?

In spite of the modern electric and digital age, there are millions of treadle sewing machines still in use around the world. Thousands of new and not-so-new treadle sewing machines are used every day in private homes and 3rd world garment and textile factories. The odds are pretty good that at some time in your life, you have worn a factory-ready-made garment that was sewn in part on a treadle sewing machine.

WHY USE A TREADLE SEWING MACHINE?

Treadle sewing machines are built to last almost forever and are actually incredibly simple devices. They lend themselves to easy home repair, service, and maintenance. A treadle sewing machine in good working order is a joy to use.The physical act of treadling can be soothing and relaxing.

The Complete Granny Miller

Many people who love to sew or quilt prefer to use only a treadle sewing machine. Often, people who sew professionally will keep a treadle machine as a backup to their electric sewing machine in the event of a power outage or a looming fitting deadline.

The needle speed on a treadle sewing machine is slower than that of an electric machine. But the slower machine speed can be a real advantage for the novice sewer. I think a treadle or hand-cranked sewing machine is the best way to teach a child or a beginner to sew. That's because it's easier for a beginner to watch their fingers and maintain control of the fabric and seam width with a treadle sewing machine.

HOW DO I GET ONE?

Modern treadle sewing machines are available new, but they can be expensive. Janome makes a fair to good modern treadle sewing machine that is supposedly popular with the Amish and other people who live without electricity. The Janome 712T treadle sewing machine uses a top-loading bobbin and has 10 utility stitches and a built-in buttonhole stitch. The last I knew, the Janome 712T is made in Taiwan and has a limited 25-year warranty.The advantage of a modern treadle sewing machine is that service, repair, bobbins, needles, and parts are readily available. The disadvantages of purchasing a modern treadle sewing machine are the lack of quality and price when compared to an older machine.

CONVERTING A MODERN SEWING MACHINE INTO A TREADLE

Necessity, and sometimes frugality, is most often the mother of invention. If you want a modern sewing machine complete with decorative stitches, many vintage sewing machines, and some modern ones, can easily be converted into a treadle or hand-cranked sewing machine. If you are handy with a screw driver, drill, hammer, wire cutters, and a jigsaw, and have a dose of creative vision and aren't a stranger at the local hardware store, then converting the right electric sewing machine may be a low-cost way for you to get a treadle or hand-cranked sewing machine. Thousands of older treadle sewing machines were converted from treadle to electric. To reverse the process is not complicated.

Many good sewing machines made during the 1920s, 1930s, 1940s, 1950s, and early 1960s have heavy grooved balance wheels that are exterior belt driven. All that is necessary to do the conversion is to simply remove the electric motor and set the sewing machine into a treadle table or base. A sturdy treadle table can be fashioned from an old treadle base with a new top. Craig's List, eBay, Facebook Marketplace, yard sales, auctions, thrift stores, Free Cycle, and plain old-fashioned asking around are all good ways to find a low-cost or no-cost sewing machine and treadle base. Sometimes, a simple classified ad in the local newspaper (old people still read newspapers) will turn up a gem of a sewing machine. Many people have old treadle sewing machines in their garages and basements that they would like to get rid of. Often, the sewing machine belonged to a beloved family member who passed away, and the family would be happy for the machine to go to someone who would appreciate it.

Depending upon the condition, such sewing machines can usually be had

The Complete Granny Miller

for $0 - $90 (2011). A word of warning: a treadle sewing machine with a base or a cabinet is heavier than a dead preacher. Be sure you bring help to load it if you plan on taking it home. If you don't have enough room for a full-size treadle sewing machine, a hand-cranked sewing machine could be a really good, low-cost non-electric solution for your sewing needs.

SOMETIMES NOTHING BEATS A REAL IRON LADY

If your heart is set on an older or antique treadle sewing machine but you don't know where to find one, or you're afraid that you can't possibly afford one, relax and be happy. Don't fret. If you're looking for an older or antique treadle sewing machine, you can probably find or assemble one to call your own. It is much easier and more affordable than you may imagine. If you know how to read and can follow directions, and aren't in too big a hurry, and don't mind some really grubby, dirty work, a beautiful old Iron Lady can be yours.

In general, there are three considerations when buying an older or antique treadle sewing machine. You must keep all of them in mind. The three considerations are:

Sewing machine head, which includes:
- bobbin type
- needle type
- feet

Base or cabinet & treadle

Availability of parts

THE SEWING MACHINE HEAD

When considering the purchase of an older or antique treadle sewing machine, the head of the sewing machine is the most important part and requires the most consideration. You will need to determine the condition of the machine head and check to see if all parts of the head are present with a visual inventory. If any parts are missing, please note which ones. When examining a sewing machine head, carefully and slowly examine the head, moving from right to left and from top to bottom.

- Does the balance or hand wheel turn, or is it frozen?
- Does the needle move?
- What is the condition of the bobbin winder? What type of bobbin?
- Who is the manufacturer? Is there a model or serial number?
- What is the shape of the base?
- Are the feed plate or plates present?
- What type of feet?
- What type of needle is used?
- Are the thread pins intact and tension disks, springs, or plates present?
- What is the condition of the steel, chrome, and decals?

The Complete Granny Miller

Sadly, nothing can really take the place of life experience when it comes to buying antique treadle sewing machines. But luckily, eBay is a great way to see lots of treadle sewing machine heads, cabinets, and treadle assembly bases. The zoom feature on eBay auction listings can give a treadle sewing machine newbie the opportunity to look up close at many different types of antique treadle sewing machine heads. Just be forewarned about eBay. Often, the seller's description is wacky and inaccurate. Complete and intact treadle sewing machine prices tend to be outrageously high on eBay and Craigslist. However, the eBay prices for sewing machine parts are reasonable. It is my favorite place to buy antique sewing machine parts, and antique sewing machine manuals are plentiful. Often, there are some real deals to be had on sewing machine heads, especially Singer, White, New Home, and Domestic. A word of advice: antique sewing machines are just like coins, guns, and rare books - the condition is everything. Just because something is old doesn't make it particularly valuable. People who are unfamiliar with treadle sewing machines tend to overprice them. At present (2011), here in Western Pennsylvania, the going auction price for a complete antique treadle sewing machine in good condition is about $55 to $125, depending upon cabinet condition and who's at the auction. Lastly, keep bobbin type and parts availability in mind when looking for an older sewing machine.

In general, bobbins are divided into 2 types: a shuttle with a spool bobbin and a modern round bobbin. Shuttles and the spool bobbins that fit into them come in different sizes and are not interchangeable. This is an important consideration when purchasing an old sewing machine. Round bobbins are a more modern system, making them easier to find and less expensive.

BUYING AN ANTIQUE TREADLE SEWINGMACHINE

When buying an old treadle sewing machine, it is wisest to look for a sewing machine that was mid-priced and popular for its time. Singer sewing machines were made by the millions and are still relatively easy to find and affordable. The Singer model 15-88 and the Singer model 66 are both good choices when looking for treadle sewing machines. The Singer 15-88 was the last sewing machine that Singer made for treadle use. Most were made in the 1930s, 1940s, and 1950s. The model 15-88 uses a modern needle, low shank feet, and has a reverse. The feed dogs can be lowered, making it a good machine for darning or free-form stipple quilting. There is a buttonhole attachment, a zigzag attachment, and a walking foot attachment for the Singer model 15-88, along with the standard hemming foot, ruffler, and other specialty feet. As of today (2011), I would expect to pay $75 to $170 for a head in good condition.

The Singer model 66 was made from 1902 to about 1960. It uses a round bobbin and a standard needle. Singer models 66s made before the early 1920s have feet and attachments that are non-standard. They have a back clamp instead of a side clamp and don't have a reverse. The Singer model 66 often has attractive and distinctive decals and is nicknamed a 'Red Eye.' As of today, I would expect to pay $40 to $90 for a Red Eye head in good condition. In fact, Craigslist and eBay are positively polluted with them.

The Complete Granny Miller

Often, it is easier and more affordable to assemble an antique sewing machinefrom parts. Old sewing machine heads tend to outlast their cabinets and bases. It's common to find a sewing machine head in good working order with a cabinet that is beyond repair. If you plan to assemble an antique treadle sewing machine from parts, it's a good idea first to find the base or cabinet that you want, then buy the sewing machine head.

Singer is my first choice for this kind of parts and pieces assembly. Singer heads will almost always fit into Singer treadle cabinets. I've never seen one yet that didn't. But measure first to be safe. Don't assume that other sewing machine brands are sewing head to treadle cabinet interchangeable. They are not. That said, sometimes cabinets and machine heads will fool you. But it is safest to stay with the same sewing machine brand. That means a Minnesota model 'A' should be moved to a Minnesota treadle cabinet. Don't take a chance trying to move it to a White, Household, or Domestic treadle cabinet. When shopping for a treadle cabinet, consider taking the machine head with you to ensure a proper fit on the spot. A good quality cabinet is scarcer than a good sewing machine head. And just so you know, it is possible to construct a new top for an old treadle assembly.

When you buy your treadle sewing machine, don't forget to hunt down an owner's manual. Many old manuals are free online, and many are available for purchase as reprints or on CDs for under $10.

CLEAN & RESTORE AN OLD TREADLE SEWINGMACHINE

Here's a list of some of the tools that you'll need to clean a sewing machine head.

- Sewing machine or household oil
- Kerosene and a tin can
- Flathead screw drivers - large & small
- Needle-nose pliers
- Tweezers
- Air in a can
- Liquid Wrench
- Extra fine steel wool
- Old Toothbrush
- Small paint brush
- Q-Tips
- Small plastic bags
- Clean lint-free rags
- Paper towels
- Car wax (optional)

A WORD OF CAUTION

Never use any type of household cleaning product on a sewing machine head. It's risky. You could destroy the decals. Ordinary household oil is usually the only product ever used to clean the exterior of an antique sewing machine head. Household oil will remove rust and layers of grime and dirt. When I bring home an old treadle sewing machine, the first thing I do is set up a neat and orderly workspace. Before I start to disassemble the sewing

The Complete Granny Miller

machine and while it's still in the cabinet, I take pictures from all angles. I photograph everything about the machine: the hand/balance wheel, the bobbin winder, the needle position, the shuttle, the tension disks, every screw, loop, or spring. I take pictures of the cabinet from every angle: the undercarriage, the skirt guard, the treadle, the pitman rod, all hinges, springs, the drawers, and the top. By keeping an extensive picture diary, I have a record of what the machine looked like before I started cleaning it. But more importantly, I have a reference for how the machine is supposed to look when I've finished cleaning and reassembling it. More than a couple of times, I've been left with an extra washer, spring, or screw from a sewing machine restoration that I couldn't figure out or remember where it was supposed to go. A photographic record will save you lots of reassembly headaches.

To clean a treadle sewing machine head, first, it must be removed from the cabinet or base. As I remove the head, I set the screws or bolts on a paper or cloth towel. Sometimes I'll letter or number the paper towel to keep track of the disassembly order and take a picture of the towel for reference. After the head is removed from the base, I will usually begin by removing the front plate, bobbin cover or covers, and any other chrome or steel pieces or fittings from the head that have screws in them. Those pieces are put on a different paper towel, which is also numbered along with the screws and photographed. Next, the tension disks are removed and placed on another numbered towel in the order in which they came off. I continue in this manner around the entire head.

Household Machine Oil Is Your Best Friend. You Can't Use Too Much

When the head is completely stripped of all removable parts, machine oil is rubbed over the entire surface of the machine head in a circular motion with fingers. Once the machine is well covered in oil, set it aside to wait for the oil to break through and penetrate the dirt and grime. Usually, after about 10 to 30 minutes, the surface can be wiped with a clean rag. Continue to massage oil into the surface of the machine head and wipe with a clean soft cloth until there is no grime or dirt left. Only clean oil should remain on the cloth. It is important to proceed gently, as too much surface abrasion will remove the decals. Clean all the fittings in the same way with household oil and rags, Q-Tips, or an old soft toothbrush. Sometimes, if I'm feeling brave, I will clean the brass or metal fittings with Formula 409 or Bar Keepers Friend and a toothbrush.

Once the exterior of the sewing head is as clean as possible, I will proceed to clean and oil the entire interior of the sewing machine head. Anything that moves or is supposed to move will get a coat of kerosene with a small brush. Let the kerosene soak into the grime and then wipe with a clean rag. Sometimes I will blow out the dirt or dust with a judicious blast of canned air - but not too much.

For a really filthy sewing machine head, I'll sometimes put it into a large covered plastic tote tub and pour 5 gallons of kerosene over it. Then leave it to soak in the kerosene in the garage for a few days. The grime will

The Complete Granny Miller

sometimes just dissolve in the tub.

Often, a couple of squirts of Liquid Wrench or any other brand of penetrating oil product will be needed if the sewing machine head action or balance wheel is stiff or frozen. The forked bars under the machine head get cleaned with a brush and kerosene, followed by oil, clean soft rags, and fine steel wool. After the last bit of grime or dirt is removed, dry the machine head with a clean rag. Apply 2 or 3 coats of a high-quality car wax and buff if desired. The sewing machine head is now ready to be reassembled and oiled. This is where the pictures and separate numbered paper towels come in handy. The parts are carefully reassembled in the order in which they came off.

CLEANING & REFINISHING THE TREADLE OR CABINET

The metal treadle assembly can be cleaned with a bucket of hot water and diluted white vinegar or Murphy's Oil Soap. Be sure to dry the metal parts after they are cleaned thoroughly. Car wax can protect and restore shine to the iron on the treadle assembly. If the treadle cabinet is in good condition, often all that will be needed is a careful cleaning with mineral spirits and a coat of paste wax. But more often than not, the wood finish will be bleached, dark, dry, cracked, stained, or peeling. Sometimes a part of the cabinet will be in need of repair. If the finish on the cabinet just looks dirty and is not in need of complete removal, I will usually try to clean it with Murphy's Oil Soap and then dry the wood with clean, soft rags. The process is repeated until the wood is dirt-free. When the wood is clean and dry, I like to apply two coats of Milsek Oil or Old English Scratch Remover over the entire cabinet and allow the product to soak in. The excess is removed with a dry, lint-free cloth.

If the cabinet finish seems hopeless and beyond the power of ordinary soap and rags, I usually remove it with paint stripper or an acetone-based furniture refinishing product like Formby's. Cabinets are best refinished by removing all hardware and working in small 6 x 6-inch sections. After the old finish has been removed, I go over the wood surface with rags and mineral spirits before I apply 3 coats of Johnson's paste wax, buffing between coats. I have also used Tung oil and Danish oil with a good result. When the base is clean and presentable, the sewing machine head is finally ready to be reinstalled. Hopefully, by the time you are done cleaning the machine head and treadle base, you will have taken my advice and found an owner's manual. The manual will give the proper threading sequence, bobbin winding procedure, needle placement, and other important information.

All you'll need to do next is to prepare yourself for a lifetime of sewing joy.

The Complete Granny Miller

BOOK 2
GARDEN FARMING

A Word from the Author

The title, 'GARDENING FARMING', is a tribute and a nod to the late author and Ohio farmer Gene Logsdon. Mr. Logsdon, who died in May 2016, was in many ways a role model for my life as a small family farmer.

Even though we never met in person, Gene's writings had a tremendous influence on my homesteading philosophy and the day-to-day workings of my farm.
He will be sorely missed.

Katherine Grossman
Plain Grove, Pennsylvania
Winter 2018

"I don't know of a better argument in favor of farming with horses than trying to start an old tractor in the wintertime."
Gene Logsdon

The Complete Granny Miller

Chapter 3

"A farm is an irregular patch of nettles bounded by short-term notes, containing a fool and his wife who didn't know enough to stay in the city." - S.J. Perelman.

Gardening Farming and Self-Supporters

The late Ohio farmer and author Gene Logsdon was the first person I ever recall to use the term 'garden farming'. I believe that the phrase is original to Mr. Logsdon. He employed the expression in order to differentiate between monoculture factory farming and the unique and small polyculture and economics involved in homesteading. Garden farming, unlike large corporate farming, is primarily a calling and not just a job or employment. Unlike large farms, garden farms are concerned with quality and are defined by low-volume and low-cost production. A garden farm does not depend upon the standard definition of economic profitably. Gene Logsdon maintained that a garden is merely a small farm. And that a farm is just a large garden. I tend to agree with him.

In the most general terms, homesteading, self-supporting, or small holding can all be defined as the home production or home manufacture of goods and services, and the avoidance of unnecessary expense or consumption outside of a household economy. Self-supporting or homesteading begins with a particular mindset that moves from a purely consumer-based lifestyle to one of increased self-reliance. Often, the move to greater household sustainably begins in a patch of dirt. By merely planting a seed, many people encounter the individual self-recognition that they have the power and the ability, at least in part, to feed themselves. This paradigm shift frequently brings about an entirely new outlook and emancipation from an unquestioning consumer mentality. A sense of economic freedom and security is born in a garden. It's because all true wealth is created by the natural and synergistic energies between man and animals, and between the sun, soil, and water. It's what agriculture is all about.

Gerald O'Hara knew it, and Scarlett O'Hara had to learn about it the hard way.

"Do you mean to tell me, Katie Scarlett O'Hara, that Tara, that land doesn't mean anything to you? Why, land is the only thing in the world worth workin' for, worth fightin' for, worth dyin' for, because it's the only thing that lasts."

– Gerald O'Hara From GONE WITH THE WIND

The basic lessons learned through simple home gardening transfer to almost everything else on a homestead or small farm. It's in the garden that we first learn the value of good planning, hard work, and perseverance. This is

The Complete Granny Miller

where we begin to understand our place and particular role in the universe. It is here that we first encounter the miracle of continuous life contained in a tiny seed. Every gardener will eventually learn to deal with failure, death, and rebirth; and the inevitability of change. Gardening teaches us patience and how to work at Nature's pace. It's in a patch of dirt that we learn to hope for a future yet unseen. Some of the most profound insights into what it actually means to be human can be found while tending the cabbages and beans.

A Garden Planner for Home Food Preservation

The estimates given in the following chart are for the vegetable production and preservation needs for a family of four for approximately 300 days. In practice, the food listed will probably last longer and extend into the following year's growing season. Quart and pint quantities are charted for convenience. These quantities can be frozen, home canned, or dehydrated. Please take into consideration that the needs of every family are different, and family food needs are constantly changing. Families with young children don't have the exact requirements as families with growing teens.

In general terms, a family of four will need approximately 600 to 725 quarts of vegetables for 300 to 365 days. Vegetable quantities are based upon the assumption of cooking from scratch three times a day, and on the assumption of a ½ cup or a 4-ounce serving size, with four servings per day. That may seem like a lot of vegetables, but it really isn't. A large pot of homemade vegetable soup can use 4 to 5 quarts of vegetables alone. White or Irish potatoes have been included in the chart. This is based upon the premise that potatoes will be served three times a week. Some families will want more than that. Some will want much less. Cabbage has been included in the chart. The amounts listed are with freezing, sauerkraut, and cold storage in mind. Tomatoes for tomato juice and tomato sauce have also been included.

You will note that some vegetables are not included in the chart. This is simply a reflection of personal food preference and the limitations of geographical location. I only eat sweet potatoes occasionally and seldom eat turnips. So, I don't grow them. If you prefer turnips, sweet potatoes, or any other vegetable not listed in the chart, consider substituting one for another. Take note that many vegetables, like celery and onions, can and will be cold-stored, frozen, or dehydrated for cooking at a later date. Also note that vegetables that cannot be preserved, like lettuce, radishes, and such, have not been listed. Only vegetables that can be preserved have been charted. There are many variables in gardening throughout the United States. The chart was written with my specific experience, growing location, and skill set in mind. Your vegetable garden may be more or less productive than mine, and you may be a more skilled or less skilled gardener than I am. Noted in the chart is the per-person foot row requirement for vegetables to be consumed throughout the active growing season. The recommendations given are generous and will probably yield more food than can be readily eaten by a family within a reasonable amount of time. This is especially true for families with small children.

Please keep in mind that food preservation needs are unique and individual. No one can determine your family's specific requirements except for you. The table is a starting point for garden planning. Extra is allowed for occasional guests at mealtimes and food gifts. Canning, freezing, dehydration, salting, fermenting, and pickling are the most common methods of home food preservation. Extra amounts of vegetables for fresh and seasonal eating have been factored into the chart, with the understanding that for 65 days, the family will eat fresh produce from the garden.

The Complete Granny Miller

VEGETABLES TO BE GROWN AND PRESERVED FOR A FAMILY OF 4

Type of Vegetable	Pounds to Raise	Quarts Pints to Preserve	Yield 100' Row	Foot Row Needed	Seasonal Foot Row Per Person	Approximate Seeds or Plants per 100' Row
Asparagus	45 lbs.	20 qt. 40 pt.	80 to 100 lbs.	75 ft.	15 crowns	75 crowns
Beets	30 lbs.	12 qt. 24 pt.	75 lbs.	55 ft.	5 ft to 10 ft.	1 oz.
Broccoli	75 lbs.	48 qt. 96 pt.	80 to 100 lbs.	100 ft.	3 to 4 plants	50 plants
Brussels Sprouts	25 lbs.	10 qt. 20 pt.	80 to 100 lbs.	35 ft.	2 to 4 plants	50 plants
Cabbage	150 lbs.	40 qt. 80 pt.	150 to 200 lbs.	100 ft.	3 to 5 plants	75 to 100 plants
Celery	75 lbs.	30 qt. 60 pt.	200 lbs.	50 ft.	5 stalks	150 plants
Carrots	120 lbs.	30 qt. 60 pt.	100 to 150 lbs.	100 ft.	5 ft. to 10 ft.	¼ oz.
Cauliflower	30 lbs.	10 qt. 20 pt.	60 to 75 lbs.	60 ft.	3 to 4 plants	50 to 60 plants
Green Beans	175 lbs.	85 qt. 170 pt.	50 lbs.	275 ft.	20 ft.	8 oz.
Wax Beans	70 lbs.	40 qt. 80 pt.	50 lbs.	135 ft.	20 ft.	8 oz.
Green Peas	100 lbs.	15 qt. 30 pt.	25 to 30 lbs.	175 ft.	20 ft.	1 lb.
Green Peppers	50 lbs.	25 qt. 50 pt.	75 to 100 lbs.	75 ft.	3 to 4 plants	50 to 75 plants
Lima Beans	90 lbs.	24 qt. 48 pt.	30 lbs.	350 ft.	25 ft.	8 oz.
Onions	125 lbs.	25 qt. 50 pt.	75 to 90 lbs.	90 ft.	2 ft. to 5 ft.	4 to 5 lbs. of sets
Potatoes	500 lbs.	400 lbs. dry store	150 lbs.	250 ft.	50 ft. to 100 ft.	10 to 11 lbs. seed potatoes cut
Pumpkin	40 lbs.	16 qt. 32 pts	100 to 150 lbs.	65 ft.	1 to 2 hills	1 oz.
Spinach	90 lbs.	35 qt. 70 pt.	40 to 60 lbs.	175 ft.	5 ft. to 10 ft.	1 oz.
Summer Squash	200 lbs.	40 qt. 80 pt.	150 to 200 lbs.	100 ft.	2 hills	1 oz.
Sweet Corn	150 lbs.	40 qt. 80 pt.	80 to 100 lbs.	200 ft.	50 ft.	4 oz.
Whole Tomatoes	350 lbs.	120 qt. 240 pt.	500 lbs.	75 ft.	3 to 5 plants	30 ground bed or caged
Tomato Juice or Sauce	300 lbs.	50 qt. 100 pt.	400 to 500 lbs.	75 ft.	n/a	30 ground bed
Winter Squash	200 lbs.	40 qt. 80 pt.	200 lbs.	135 ft.	2 to 3 hills	1 oz.

Katherine M. Grossman © 2009 GRANNY MILLER

The Complete Granny Miller

The Difference Between Hybrid and Non-Hybrid Seed

New gardeners are sometimes confused by the terms hybrid, non-hybrid, or heirloom when shopping for vegetable or flower seeds. Here's a quick primer.

Hybrid Seed

Hybrids are the result of breeding two distinct varieties of plant or animal. The offspring of such a mating produce a new variety of the same species that incorporates specific desirable characteristics of both parents. There are many good reasons for using and breeding hybrid plants or animals.

Hybrids, whether they are plant or animal, are usually more vigorous and healthier than their parents. Fruits and vegetables from hybrids will often outproduce their parents with more size and shape uniformity. Many hybrids show greater resistance to pests and disease, and often will ripen sooner or set fruit earlier. This increase in hardiness and productivity is known as 'hybrid vigor'. The first generation of a hybrid cross is referred to as the first filial generation or F1 for short. Hybrids don't necessary have only two parents. Many hybrid animals and seeds result from the cross-mating of three or four different individuals.

Seeds or offspring from hybrid vegetables or animals will not produce the same type of plants or offspring as the parent. Hybrid seed saved from hybrid parents at best will come true to the parent less than 30% of the time. Seed from some hybrid plants can even be sterile. The nonperformance of saved hybrid seed or F2 generation seed is an important economic and environmental adaptation consideration for those who wish to save seed from year to year. F2 seed usually will have lost the hybrid vigor from the true hybrid parents. If planted the following year, the seed will begin to revert and exhibit undesirable characteristics that may be part of the genetic makeup of one or both of the parents. Hybrid seeds must be purchased every year, and they are not cheap. Because hybrid plants, by their nature, cannot produce a dependable and predictable seed, they cannot, over many generations, adapt to specific environmental needs or situations.

Genetically Modified Seed

Another way that seed can be made hybrid is by genetic modification. Such seed is often referred to as genetically modified or 'GM' seed. GM seed is created in a laboratory by introducing the DNA from one species or organism into another. Genetically modified seed has attached proprietary rights and is known to contaminate open-pollinated crops. This cross-contamination has resulted in several high-profile lawsuits. In some cases, genetically modified seed has what is known as a 'terminator gene'. This is a gene sequence to make the resulting F1 seed sterile purposefully. The terminator gene developed by Delta and Pine and now owned by Monsanto was never approved for sale in any crop. There are no seeds available for sale that contain this gene sequence. The Center for Environmental Risk Assessment maintains a database of genetically modified plants that have regulatory approval for sale in the US, though not all are commercially available. The use of genetically modified seed and the cross-contamination of crops by genetically modified organisms has become a worldwide hot-button agricultural, economic, health,and environmental issue. The use of GM seed is one of the reasons for mass suicide among small farmers in India.

Non-Hybrid and Open-Pollinated Seed

Non-hybrid seed is often known as 'open-pollinated' or OP. Open-pollinated plants are crossbred varieties that one gardener has often passed down to another gardener for many generations. The seeds from these crossbred plants are quite stable. These seeds produce plants that grow true to the parent with only a few sports or genetic mutations. Plants from open-pollinated seed over many generations can adapt to a specific local environment or growing conditions. Non-hybrid seed can be collected every year from open-pollinated plants and stored for the next growing season. When collecting seed for future use, only seed from the most superior plants must be saved. Unlike hybrid seed, seed saving from open-pollinated plants is free and incurs no yearly cost. This is an important economic consideration for gardeners and small traditional farmers. Keep in mind that if you are growing non-hybrid, open-pollinated plants for their seed, they must be kept well away from other plants of different varieties so that they don't cross-pollinate. It's the wind rather than insects that usually carries pollen. Distance is the most effective tool and insurance against cross-pollination. A distance of 250 ft. to 300 ft. between different types or varieties of plants will ensure that the seed comes true to the parent. A distance of 600 ft. to 700 ft. will give complete isolation and is only used for scientific or plant breeding purposes.

Heirloom or Heritage Seed

Heirloom or 'heritage seed' is simply a type of non-hybrid seed. There is no general agreement on the use of the terms heirloom or heritage when describing seeds or plants. Heirloom seeds are always open-pollinated and non-hybrid.

The term is commonly used to describe seed or plant varieties that were grown prior to WWII. This was a time before the beginnings of big industrial agriculture or the Green Revolution. Some growers feel that the term heirloom should only be applied to seeds or plants that have been passed down in one family to the following generation of family gardeners. Many people believe that heirloom seeds always produce a tastier, superior, and more nutritious fruit or vegetable. This is simply not true. It depends upon the plant.

Modern hybrid sweet corn is my favorite example of this ill-informed Internet myth. It's my opinion that any variety of modern sweet corn is far superior to the common heirloom sweet corn varieties, such as Golden Bantam or Sunshine. Every year, I plant a couple of different types of sweet corn in my vegetable garden. Bodacious and Silver Queen are my favorites. Old heirloom varieties of corn, such as Early Butler or Country Gentleman, are best used as animal feed.

I particularly love and appreciate the many different varieties of heirloom tomatoes. Amish Paste, Ox Heart, Abe Lincoln, Brandywine, and Old German Striped are some of my favorite heirloom tomatoes. There's no

The Complete Granny Miller

question that open-pollinated heritage and heirloom seeds help to ensure worldwide plant diversity. In fact, plant and animal diversity and food crop genetics are the cornerstone of worldwide food security. As of late, this issue has become more critical than ever.

Which brings me to …

Do you know about the Svalbard Seed Bank? The reality of limited plant genetics is one of the reasons for the Svalbard Global Seed Bank, which has become fodder for modern-day apocalyptic global famine scenarios and conspiracy theorists.

The Svalbard Seed Bank is also known as the Doomsday Vault. The vault is located 810 miles from the North Pole on the Norwegian island of Spitsbergen and houses seeds from every continent. The concept of the seed bank was to provide insurance against the loss of plant genetics in the case of catastrophic regional or global crises. The Svalbard Seed Bank cost over $9,000,000 in U.S. dollars to build and was funded entirely by Norway. The government of Norway and the Global Crop Diversity Trust provide for the day-to-day operational costs. The Global Crop Diversity Trust is an independent international organization. The seed vault has sci-fi state-of-the-art security systems, but no permanent staff.

"As Soon As the Soil Can Be Worked"

Sometimes new gardeners need help interpreting and deciphering the back of a seed packet or a garden catalog. It seems like the information or directions should be forthright and easy to understand, but that's not always the case. There are a few terms and concepts that may be encountered in garden catalogs and on seed packets that can leave the novice gardener befuddled and in need of a more detailed explanation. The statement, "as soon as the soil can be worked," is often found on the back of seed packets for early spring flowers or vegetables. Plants like peas, pansies, and other early spring plants are typically directly sown into the garden. Sometimes the expression can confuse an inexperienced gardener.

"As soon as the soil can be worked"

The saying is a description of soil conditions and not about any particular day on the calendar. Soil can be worked and seeds planted when the soil is no longer wet or sticky. When your garden can be worked will depend upon what type of soil you have. Keep in mind that your soil may be different from that of your neighbor. Just because your neighbor's garden is being planted doesn't mean you should plant yours too. Sandy soil can be planted sooner than clay-type soils. Whatever you do, don't disturb the ground before it is ready, or you'll ruin the soil structure.

To test if your soil is ready for planting, grab a fist full of soil into your

The Complete Granny Miller

hand and squeeze it together into a semi-conical or oblong shape. If the soil sticks and clumps together when squeezed with just a little pressure, the ground is too wet to plant. You'll need to wait for the soil to dry out more. The ground is ready to be planted when the soil is a little crumbly and will readily fall apart after being squeezed in your hand. Sometimes the soil can be marginal ready and it's a flip of a coin whether or not to chance planting. With borderline conditions, whether or not to plant would be determined solely by the weather forecast. If the weather looked like it was going to be warm and dry for the next few days, it's probably safe to plant. But if cool and rainy weather is on the way, absolutely do not plant that day. Seeds planted under cool, wet conditions will rot in the ground and need to be replanted.

With gardening as with most things in life, no harm ever comes from waiting and working with Nature. If you jump the gun with Spring Fever and disturb the soil too early, you'll pay the devil to make it fit again.

How to Test Seeds for Germination Rate

Germination rate is the term used to describe the percentage of seed from a particular plant species that will emerge when given the right conditions. The germination rate is vital information to know ahead of time and helps the home gardener or farmer determine how thickly or thinly any given seed needs to be sown. Germination rate is also a consideration whenever you are ordering seeds from a catalog or if you have leftover seeds from a previous year that you want to use. Seeds will not last indefinitely, and some seeds are more time sensitive than others. Keeping all types of seed cool and dry will go a long way in helping to preserve seed viability. Most vegetable seeds will last about 3 years, and some can last as long as 6 years.

Corn, onions, leeks, chives, green peppers, parsley, and sometimes parsnip seeds are all seeds that don't store well. They should be tested every year before planting if they were held over from the prior year.

Beans, peas, beets, cabbage, Brussels sprouts, broccoli, kale, carrots, eggplant, squash, pumpkins, tomatoes, basil, sage, rue, borage, anise, dill, oregano, and most herbs have a seed life of between 3 to 4 years.

Spinach, lettuce, cucumbers, and most melons have a seed life of about 5 years.

It's always a good idea to test the germination rate if the seed is much older than 2 years old or wasn't stored under ideal conditions.

For most common garden vegetables or flowers, seed germination rate is easy to test at home. To test seeds for viability, I start by counting out 20 seeds and placing them on a small plate lined with a wet paper towel, or inside a wet paper towel with a small amount of potting soil sprinkled on it. The paper towel or lined plate is next placed inside a plastic bag and sealed. The sealed test seeds in the plastic bag are then left on top of the refrigerator or in a warm room for about a week to 10 days. Germination time will depend upon the type of seed. Some types of seeds take longer than others to germinate, and some seeds require total darkness for germination. Once

The Complete Granny Miller

the seeds begin to germinate, I count them to determine the percentage. By using 20 seeds for testing, I'm able to get the percentage in 5% increments. To evaluate germination rates, the following percentages are helpful.

- 100% - Great! You can't do any better than that.
- 90% - Excellent. You can count on the planting rate information supplied with the seed.
- 80% - Good. You shouldn't have too many problems as long as environmental conditions are favorable.
- 70% to 60% - Poor. If you want to use the seed better, sow it thick.

Starting Vegetable Plants from Seed

I don't start too many vegetable plants indoors. That's because houses are great for humans but are unnatural for vegetables. From a plant's point of view, most homes are too dark, dry, and hot. Vegetable plants tend to do better when started in a cold frame or a hotbed. The only plants I routinely start indoors are tomatoes, green peppers, and sometimes melons.

About eight to ten weeks before the last expected killing frost, I sow seeds for tomatoes and green peppers. As an advocate for planting by the moon, I try, if possible, to pick a favorable day for sowing seeds.

When starting seeds, I usually use a standard terracotta flowerpot and sterilized potting soil with a slow time-release fertilizer. For really special seeds, sometimes I'll sow them in individual eggshell pots or small peat pots. Because I most often use old pots from the garden, I first make sure that the pots are well scrubbed with hot soapy water before sowing the seed. It's important to kill bacteria or fungus that could have been held over from the prior year. Young seedlings grown indoors are susceptible to damping off. Cleanliness and good sanitation will go a long way in preventing plant troubles. For tomato and pepper seeds, I fill pots with sterile potting mix to within 2 inches of the top, and tamp the soil down firmly. The seed is next lightly broadcast into the pot and gently sprayed or misted with warm water to moisten it. Potting soil mix is then sprinkled evenly about ¼ inch deep over the seed. The soil is then tamped down firmly again, ensuring the seed makes good contact with the moist soil. Seeds are somewhat forgiving, and they will take being planted deeper. But it's usually better not to over-bury the seed if you can help it. As a rule of thumb, most seeds need to be planted about three or four times as deep as they are wide. When planting different varieties of the same plant, I clip the seed packet onto the pot with a clothespin or rubber band before putting the pot into a large plastic bag.

Most newly sown vegetable seeds don't require sunlight to germinate. They only need warmth. In my house, the most dependably warm place is on top of the refrigerator. So that's where the seed pots go until the seedlings begin to emerge. Once the seedlings emerge, I remove the plastic bag so that they can breathe. The seedlings are then moved to a sunny window ledge during the day. At night, the seedlings are moved away from the window.

Nighttime windows can become too cold and may damage or stunt young plants. While the plants are young, it's important not to overwater them. Over-watering encourages 'damping off'. Damping off is a plant condition cause by several different types of pathogens that weaken young plants and seedlings. It can even prevent seeds from germinating.

Once the seedlings start to compete with each other in the seed pot, I will transplant them into individual peat pots or small plastic pots.

When the weather begins to settle with the coming spring, young seedlings are 'hardened off'. This is done by setting them outdoors for brief periods during daylight hours. The hardening off process gives young plants time to adjust to the garden. To harden off seedlings, pick a warm, sheltered location well away from wind and full sun. After about a week or 10 days of hardening off, if the weather is settled, the new bedding plants can be moved to the garden.

One way to tell if the weather has truly settled enough for cold-sensitive plants to be moved to the garden is by the size of oak tree leaves. If the leaves on an oak tree are as big as a mouse ear, it's safe to plant.

Egg Shell Seeding Pots

Did you know that back in the days before commercial peat pots and plastic trays, people used half of an eggshell to start seedlings? That's right. An eggshell is a good container to sow a seed in. Egg shells are small, lightweight, porous, earth-friendly, and readily available. In fact, eggshells are free if you own chickens. Small slow slow-growing seeds like peppers or onions work best. Here's the way to make and use eggshells for seed pots.

Crack a raw egg high up on the small end. I use the dull edge of a dinner knife to make the crack. Empty the contents of the egg and rinse out the shell. Poke two or three small holes in the bottom of the shell with a sharp, pointed tool. A metal shish kabob skewer or a large darning needle is suitable for this. A large nail works well, too. Fill the half shell with seed starter mix and sow a seed. Bury the eggshell halfway in a container filled with sand. The Styrofoam top of an egg carton works well for a sand container. Cut the egg carton on the fold line to fit on a windowsill. Water the sand and then cover the container with plastic. Set the container in a warm place until germination.

After the seeds germinate, remove the plastic and move the eggshells to bright light. Always water the sand in the container instead of the eggshell. Eggshells are porous, and the holes poked into the bottom allow water to wick up to the plant.

When it's time to plant the seedlings in the garden, gently crush the eggshells or remove the seedlings without disturbing the root systems. Keep in mind that most seeds sown indoors without bottom heat are started 4 to 6 weeks before the last frost date. Not much good comes from starting seeds too early. Plants started too early tend to become weak and spindly.

The Complete Granny Miller

Soak Hard Seeds for Easier Germination

Sometimes hard seeds need a little help to germinate. Morning glories, sweet peas, or moon flower seeds are examples of seeds with a hard exterior. If you soak seeds with a hard coating for 24 hours before you plant them, they will germinate more easily. Adding a little milk or vinegar to the soaking water helps to break down the hard seed coating. Another trick is to nick the seed with a nail file or sharp knife before it is soaked.

Choosing a Vegetable Garden Location

Often, new gardeners are confused by the abundance of garden information and advice. There are as many opinions about vegetable gardens as there are gardeners. But the happy fact is, vegetable gardening just isn't that complicated. All that is needed to grow a good and productive vegetable garden is a dose of old-fashioned common sense, planning, some hard work, and a little knowledge of how plants grow.

Even if you don't have lots of ground to work with, container gardening is one way to grow more food than you'd imagine in a limited space. Container gardening confined to a high-rise balcony or a patio deck will benefit from the same recommendations as a raised bed garden, a square foot garden, or traditional in-ground row planting.

One of the key factors for a flourishing and thriving vegetable garden is proper site location. In general, the only requirements for a good vegetable garden location are full sun, level ground, proper soil pH, and access to water.

Sun

The first and most important requirement for any vegetable garden is strong, full sunlight. By that I mean at least 6 to 12 hours of full sun every day, with little to no shade. Most vegetables require full, uninterrupted sunlight in order to be productive and to stay healthy. Inadequate sunlight will result in spindly leggy plants that are slow to flower, slow to fruit, and subject to disease and garden pests.

When selecting a garden site location, it's wise to invest some time observing how the sun moves throughout the day for any given spot. If possible, watch the potential site for a full year. That's because the arc of the sun changes with the seasons, and the angle of a summer sun is different from that of a winter sun. Pay extra attention to nearby trees and buildings when selecting a new garden site. Sometimes a potential location will have a close tree, fence, or building nearby, which may cause unseen summer shade or drainage problems. Here's an example of what I mean.

The eastern or northeastern exposure of a location may have the morning sun, but by the afternoon, the location is in shade. That's not good. The same problem can occur with a full western exposure but in reverse. With a western exposure, there's no morning sun to help dry the dew off the plants; only a full burning afternoon sun. A northern exposure will not have adequate sun for vegetables and will be in shade throughout the day. A southern exposure is ideal because it provides no shade once the sun rises. Once a full sun location has been established, it's worth mentioning that the way a garden is oriented in relation to the sun is a factor in maintaining full sun for all plants throughout the day. A rectangular shape, north-to-south orientation, is the most successful vegetable garden configuration. Even with correct sun orientation, it's important to plant seeds and bedding plants in a way to maintain full sun exposure for all vegetable plants. The way that vegetable seeds or bedding plants are sown or arranged in the garden can make a difference in how much sun individual plants receive throughout the day.

Whether you are using a raised bed system or row planting, the actual planting of seeds or bedding plants should be thought out ahead of time. That's because not all vegetable plants are the same height. Once the tall vegetables in a garden attain any height at all, they will begin to cast a shadow on the low-growing vegetables and shade them from the sun if not appropriately oriented. Tall-growing vegetables like corn or climbing beans should be planted at the top or north end of the garden. Low-growing vegetables like beets, carrots, or lettuce should be planted at the south end or bottom of the garden. A gardener can use this information to their advantage for some vegetables. Certain vegetables like lettuce, peas, beets, or spinach will tolerate partial sun, especially during the hottest part of the summer. With a bit of shade, vegetables like radishes or beets will not have to be watered as often and will not tend to wilt from heat.

Level Ground If Possible

A level location is desirable when planning or planting a vegetable garden. If there's too much slope to the ground, water and soil will wash down and drain away from the high part of the garden. Water will tend to pool and settle at the low end of the garden. That's a bad situation for plants located at both ends of the vegetable garden. Plants on the high end of the garden will not receive enough water, and the soil will tend to erode from the roots. Plants located at the low end of the garden will often be sitting in mud or in standing water for an extended period. Not a good garden situation if it can be avoided.

If there is no other choice but to use a hillside or non-level ground, orient the rows so that they run horizontal across the hill and not up and down vertically, no matter what the sun orientation or the direction the garden faces.

Proper Soil pH

The following requirement for choosing a successful vegetable garden location is proper soil pH. Soil pH is a fundamental part of all gardening. Even though gardeners may not be able to choose the soil or dirt in their gardens and backyards, they can work with what they have. Soil pH testing kits or pH meters are inexpensive and available at home garden centers and online. Garden soil pH should be tested once the garden site location has been selected, and again after the soil has been prepared when the soil is dry.

Soil with a pH above 7 is considered to be an alkaline soil. A soil with a pH

The Complete Granny Miller

lower than 7 is considered to be an acid soil. In general, soils in moist climates are acidic, and soils in dry climates tend to be alkaline. The addition of lime can amend a soil that is too acidic. The addition of sulfur can correct soil that is too alkaline.

My garden soil tends to be acidic, and I use ordinary hardwood ashes to amend it. Common garden vegetables that prefer a slightly acid soil are corn, potatoes, tomatoes, radishes, and sweet potatoes. Vegetables that prefer a slightly alkaline or 'sweet' soil are beets, broccoli, beans, cabbage, lettuce, summer squash, pumpkins, asparagus, and cucumbers.

Water

The last consideration in locating a vegetable garden is a convenient source of water. If you are feeling lucky, you can sit back and let Mother Nature water your vegetable garden whenever it rains. But more often than not, you'll be sorry if you take a chance with her. Most people will want to water their gardens on a regular basis. If possible, try to have a source of water or access to a garden hose nearby. Carrying heavy buckets of water over a long distance during the summer gets old fast, especially when the bugs are biting or during a merciless heat wave.

By considering the sun's location and movement, water accessibility, ground level, and soil pH, you'll be well on your way to achieving vegetable gardening success.

Reasons to Use a Broadfork in Your Garden

A broadfork is a low-tech garden tool. It's sometimes called a 'U-bar' because of the square shape. A broadfork is used primarily to break open the earth to permit aeration and deeper soil penetration. It allows for better plant root development without an undue disturbance to the basic soil structure. The proper use of a broadfork will not destroy earthworms or their tunnels, nor the beneficial flora and fauna in the soil. It does not collapse or invert the soil like a rototiller will. A broadfork can work and loosen the soil at more than twice the depth of standard four or five-prong garden forks. It can loosen the soil much deeper than a rototiller can. A broadfork is easier to use than a regular garden fork and can cover a greater amount of garden space in much less time. That's because a garden fork requires the gardener to use their shoulders, arms, and back. Instead, a broadfork uses a natural back-and-forth rocking motion. Most of the work is accomplished by the weight of the gardener standing across the tines of the broadfork. The deeper the soil is loosened, the better plant roots can develop.

No other garden tool can loosen soil or hardpan to the depth that a broadfork can. Unlike a rototiller, a broadfork requires no pushing, no fossil fuel, and is ideal for raised bed gardens. It's also kind to families on a budget. A broadfork can be used to mix nutrients into the soil in early spring or fall. It is helpful for digging root crops like potatoes. For most modest-sized family gardens, the only real garden tools needed are a good garden rake, a shovel, a hoe, and a broadfork.

The Complete Granny Miller

How to Plant & Grow Garlic

Fresh garlic is an indispensable addition to my cooking. I use a lot of it throughout the year. So, I grow a lot of it.

Every year, I save seed garlic from the harvest in July to replant in October for the following year.

'Seed garlic' is the term for a bulb of garlic in which the cloves have been separated. The individual cloves are the seed. When the cloves are joined together, the garlic is called a bulb. Garlic belongs to the genus Allium and the species sativum. Garlic is divided into two distinct types: hard-neck and soft-neck.

The hard-neck type of garlic does well in northern climates: a cold winter followed by a wet spring. There are three main types of hard-neck garlic: Rocambole, Purple Stripe, and Porcelain.

Soft-neck garlic does well in warm climates and is the type you'll find in most grocery stores. Soft-neck garlic ships well. The two main types of soft-neck garlic are Artichoke and Silverskins.

Garlic is easy to grow. And if you love it like I do, it deserves a place in your garden.

The most important tip for growing great garlic is to start with the correct variety for your area. Not all varieties will grow everywhere with success. Select plump, disease-free cloves for planting. Rich, well-tilled soil is an advantage when growing garlic. Good, loamy soil always gives a bigger bulb that produces a thicker stem. A thick stem is an indication of the size and health of the cloves.

October and November are the best planting times in my area to plant garlic. I try to wait until after a couple of hard frosts before planting. Garlic tends to heave up out of the ground if planted too early.

For those who like to garden by the moon, plant garlic in the first or second quarter when the moon is in the sign of Scorpio or Sagittarius. Harvest garlic when the moon is in a dry and barren sign like Aries, Leo, or Sagittarius.

Planting

Choose a location with good drainage and full sun. Dig a long, shallow trench about 2 inches deep and mark the two ends of the row. It's essential to mark the trench row so you'll be able to find the garlic shoots in the spring. Garlic doesn't always send up green sprouts in the fall, and you may forget where it was planted.

Place a nice-sized clove, wide end down and pointed end up, snugly into the trench so that it makes good soil contact. Space the cloves about 4 inches apart. Cover the cloves with soil and tamp the soil down firmly. Unlike some other plants, rows of garlic may be spaced quite close. I have success with a 12-inch spacing of rows.

In the spring, keep the garlic well-weeded and apply mulch if you want. The mulch will help to retain soil moisture and discourage weeds. I use straw for mulch because it is readily available to me. But you can use anything. Old newspapers, black plastic mulch, or grocery bags will work fine as long as you weigh down and anchor the sides with soil. If the weather is dry during the spring or the summer, keep the garlic well-watered. But don't allow it to become waterlogged, or it will rot.

Harvest
Garlic is ready to harvest when the top of the plant starts to turn brown, die back, and topple over. The time that this occurs varies from year to year. For my location, it's usually at the end of July or the beginning of August. The best way to test for harvest readiness is to pull up one garlic plant and check the sheaths that surround the bulb. The sheath is the layered paper-like covering that encases the cloves. A covering of two or three layers of sheath is ideal for garlic.

Harvest garlic by pulling it up carefully or by using a garden fork. The bulbs will have soil clinging to the roots. Spray off the soil from the bulbs with a garden hose. Then place the bulbs upon a heavy wire rack or window screen to dry out in the sun for a day or two. After a couple of days, move the garlic to a dry, sheltered location for a couple of weeks to finish the curing process.

When the bulbs are completely dry, cut off the scraggly roots. Store garlic by braiding the stems together or by tying them tightly into bunches with twine. Garlic braids are fabulous hanging in a kitchen. Garlic keeps best in a dry location at a temperature of between 50°F to 70°F.

When the fall comes around again, select the biggest and most perfect bulbs for seed and start the process all over again.

How to Plant Asparagus

Asparagus is a perennial vegetable. That means it will come up every year, and you only have to plant it once. Happily, it's also one of the first vegetables to appear in the garden in spring. The stalks of asparagus grow upward from the roots or crowns in mid-spring. By mid-summer, asparagus will produce light, wispy ferns that stay on the plant until the fall. After a few hard frosts, the tops of the asparagus will die back. During the winter, asparagus lies dormant in the earth, waiting for spring to start the growth cycle again. When properly planted and cared for, asparagus will dependably produce every year. A well-tended asparagus bed will last between 4 to 15 years and sometimes much longer. A lot depends upon the variety that is planted, the cultivation practices, and the soil conditions. I've

61

heard of asparagus beds that were well over 100 years old.

In the home garden, most asparagus is grown from crowns and not from seeds. In botany, crowns are the section of a seed plant where the root of a plant joins the stem. Planting asparagus from seed is time-consuming due to the special care that the young plants require. When planting from seed, asparagus will not be ready for a minimal harvest for at least 3 or 4 years. The modern hybrid varieties of asparagus are best for most home gardens and small farms. Jersey Knight, Jersey King, and Jersey Supreme are all good, dependable hybrids. They produce all male plants and waste no energy producing seed. Asparagus planted from crowns will begin to produce in the second year after planting and can be lightly harvested. Some gardeners will wait to harvest until the third year after planting to ensure that the plants have fully matured.

A single asparagus crown will produce about ½ lb. to 1 lb. of spears after three or four seasons. For garden planning purposes, 12 to 18 crowns per person for seasonal consumption is considered adequate for most families. A 100 ft. garden row will yield approximately 80 to 100 lbs. Plan to plant twice as much if you intend to freeze or can asparagus.

Asparagus should be planted after the garden soil has warmed up to about 50°F. No good comes from planting in wet or soggy soil. Asparagus will not grow until the ground has warmed. Planting in wet, cold conditions increases the chance of crown rot. Ideally, asparagus should be planted in a separate garden bed. If that is not possible, plant them on the west or north side of the central garden so that they will not shade other plants during the summer.

For gardeners who like to plant by the moon, asparagus should be planted during an increasing or waxing moon, when the moon is in the sign of Cancer. If a Cancer day cannot be chosen, Scorpio is the next best day, and then Pisces.

When I was a young gardener, I planted asparagus beds the old-fashioned, traditional way. The traditional way of planting is time-consuming and back-breaking. And if you have hundreds of feet to plant, it really isn't practical. With the traditional method of planting, a 9 to 12-inch deep trench is first dug. Then each root or crown is laid on top of a small saddle of dirt. The crowns are next covered with about 3 to 4 inches of soil. Slowly throughout the summer, the trench is filled in with more dirt as the asparagus grows and develops.

But about 8 years ago, I discovered a better way to plant asparagus. The new way to plant is so much faster that it lends itself to small commercial production. Since that time, both my asparagus patch and my chronic springtime backache have improved. The new planting method is so much faster and easier than the traditional way that I'll never plant asparagus any other way again. Here's how to do it.

First, dig a long ditch or row trench about 6 to 8 inches deep. In the bottom of the trench, sprinkle 1 lb. of 0-46-0 (triple superphosphate) or 2 lbs. of 0-20-0 (superphosphate) fertilizer for every 100 ft. of row. By doing this, phosphorus is made immediately available to the crowns. If you skip this step, your asparagus will not produce as well, and the stalks will be spindly

The Complete Granny Miller

and weak. Now comes the unbelievable easy part. Walk along the row and toss the crowns into the ditch on top of the fertilizer, keeping the crowns about 9 inches to 12 inches apart. It doesn't matter how the crowns land in the ditch or if they make direct contact with the fertilizer. The fertilizer will not burn the crowns.

Next, wholly but lightly fill in the ditch. But do not pack the soil down too hard. If you compact the dirt too much, the emerging spears will struggle to reach the surface and the sun. We want to make it easy for them. Keep the ground moist and well-watered, and new growth should be seen within 2 to 3 weeks. If you are going to plant more than one row, maintain wide rows with a spacing of 4 ft. from center.

Keep the young plants well-weeded. In the fall, don't be in a hurry to cut back the top ferns. Instead, allow the asparagus to die back from frost naturally. If too much foliage is removed from the top of the plant, the roots are prevented from storing as much energy and food as possible. This interferes with new stalk production the following spring. In the past, I have always managed weeds by the application of salt. But that method of weed control is no longer recommended. The salt will not hurt the asparagus. Asparagus is native to coastal areas and loves salt. But the salt can alter the soil and harm the surrounding vegetable plants. Presently, the standard practice for weed control is to burn off the weeds with a small controlled fire in the early spring or late fall, before the spears begin to emerge.

In the spring, once the soil temperature has warmed up to around 50° F, asparagus spears will begin to appear. To harvest asparagus spears, snap them off at the base when they are about 6 inches to 9 inches tall. Don't cut them below the ground line, or you might injure other buds that have yet to emerge.

Here's what I do to keep my asparagus bed healthy.

Every other year or so, I top dress my asparagus patch with some type of livestock manure while there is still snow on the ground. As the snow melts, the nutrients from the manure are broken down and pulled into the earth. A top dressing with hardwood ashes will also keep the plants green and vigorous. Asparagus uses a lot of potassium, and hardwood ashes are rich in it. Superphosphate fertilizer is only applied at planting time and never afterward.

And just in case you're too shy to ask. Asparagus causes a characteristic pungent odor in some people's urine shortly after they consume it. The odor can be noticed within 15 to 20 minutes of eating asparagus and is known as 'asparagus urine'. Asparagus urine studies were conducted a while back. It seems that almost everyone may produce the chemical components that make for smelly asparagus urine. But not everyone can smell it. Science has come to no agreement on this issue. What researchers can agree upon is that the ability to smell the odor in asparagus urine and the ability to produce asparagus urine seem to have a genetic component.

So, if you don't have the gene to smell asparagus urine, consider yourself lucky. Enough said.

The Complete Granny Miller

How to Propagate Plants by Stem Cuttings

One of the fastest and easiest ways to reproduce certain types of plants is by a method known as 'stem cuttings', 'truncheon(s)', 'striking', or 'cloning'. It's an asexual method of plant propagation. It usually requires only a small leafy section from the stem of a parent plant. When the stem section from the parent plant is appropriately prepared and subject to the right conditions, it will grow roots and will become an independent plant and a clone of the parent.

Plant propagation by stem cuttings is a foundation gardening skill that is fun, economical, and easy to learn. Once you acquire the art and skill to clone a plant from a stem cutting, a large part of the plant world becomes your playground. You might even find yourself carrying a pocket knife and a plastic bag everywhere you go. What's more, if you have a green thumb and are so inclined, the propagation of plants by stem cuttings can become the underpinning for a lucrative home-based business; however, a few words of caution.

Some plants are patent-protected. So, take care not to infringe upon or break patent laws by offering a patented plant for sale. If a patented plant is propagated for your use and not for commercial gain, there is usually no patent protection problem or issue.

What follows is a table of some of the herbs, flowers, and shrubs that stem cuttings can easily propagate. The table is by no means a complete list of what can be propagated via stem cuttings. Hundreds of different varieties of plants, herbs, shrubs, and trees can be propagated by the home gardener. I would encourage you to visit gardening websites or forums to see the various other types of plants that other gardeners have reproduced in their gardens.

Roses	Rosemary	Rose of Sharon
English Ivy	Mint	African Violets
Chrysanthemums	Lavender	Poinsettia
Blueberry	Geraniums	Gardenia
Elderberry	Thyme	Forsythia
Holly	Yew	Lemon Verbena
Wisteria	Lilac	Jasmine
Bay Laurel	Mock Orange	Rhododendron

Here's What You'll Need

When it comes to reproducing new plants from stem cuttings, not too much is needed in the way of special equipment. Most of what is required can be found around the homestead or in the garden shed.

The method I use employs:

- Parent plant
- Months between May to September
- Clean pot with good drainage
- Sterile potting soil

The Complete Granny Miller

- Wide-mouth canning jar
- Small knife
- Garden pruners
- Watering can
- Semi-shady location
- Honey
- Patience

Choosing the Parent Plant

Basically, there are four distinct types of cuttings or truncheons because there are different types of plants. The four types are: herbaceous cuttings, softwood cuttings, semi-hardwood cuttings, and hardwood cuttings. The type of cutting and the timing for the propagation depend upon the plant that you choose to propagate. That's because not all plants have the same growing habits or needs. A simple comparison, for example: A spider plant, which is an indoor herbaceous plant, will start to grow roots much faster than a boxwood, which is a semi-hardwood outdoor shrub. The spider plant doesn't favor a particular season to be reproduced by cloning. But the boxwood most certainly does. An understanding of the growth habits of the particular plant you want to reproduce helps to ensure greater propagation success.

I have found the easiest plants to reproduce are usually herbaceous plants and softwood shrubs. Many houseplants and some outdoor plants are quickly propagated by simply cutting a leaf or a branch and inserting it into a jar of water.

African violets, pussy willow, mint, begonia, coleus, roses, philodendron, sweet basil,and many more plants can be started this way. The downside of rooting a stem cutting in water is that the roots are often brittle and will break or degrade when planted into soil. A much better way to start a stem cutting is by properly preparing the stem cutting and planting it in sterile soil.

One Way to Root Stem Cuttings

Here's a quick way to start blueberry bushes, pussy willow, trumpet vines, boxwood, juniper, jasmine, and hundreds of other plants.But be forewarned. Not all plants will have the same success rate. So, it's essential to be prepared for failure and some losses. But also, be prepared to be pleasantly surprised. More than once, I've had a 100% success rate. A lot depends upon the plant variety, weather conditions, and a gardener's attentiveness. To begin propagation, collect and prepare clean pots filled with sterile soil ahead of time. Gather together the same number of 2-quart Wide Mouth Mason jars. If you don't have 2-quart jars, a 1-quart jar will work. But understand that when using a 1-quart jar, the cuttings will have to be shorter, and won't have the extra headroom that a 2-quart jar provides. Stem cuttings need humidity and moisture in order to grow roots. The Mason jars are placed over the cuttings and act like little greenhouses. They are removed after the cutting has begun to grow its roots.

I usually collect stem pieces during the morning hours of early summer.

The Complete Granny Miller

That's because stem cuttings with actively growing softwood are easily found. Softwood is a horticultural term used to describe the stage of growth on a deciduous woody plant. Softwood is not the soft, leafy growth at the top of the plant, nor is it the stable, inflexible woody growth lower down. Softwood is a flexible plant growth that is easy to snap, lying between the two. Early summer is perfect for that type of growth.

Morning hours will most often find plant stems turgid and well hydrated. Actively growing softwood is quite firm. It will have mature leaves and leaves that are still immature. When taking stem cuttings, I aim to find sections from the bush or plant that have offshoots or a 'V' type notch in the stem. The section on the stem where the offshoot branch or 'V' notch is located contains specialized cells or growth nodes. These nodes will readily root the stem section once it is adequately prepared.

To begin the cloning process, cut about a 6-inch to 12-inch section from the parent plant. The section must have healthy leaves and no flowers or flower buds. That's because the plant's energy should go towards root formation and not to flower production. When collecting the stem cuttings, place and hold the cuttings in a bucket of water in the shade while you work. The stems mustn't dry out or become stressed in any way. When preparing the cuttings for planting, it's also best to work in the shade.

For each stem cutting, strip away about two-thirds of the leaves and remove all thorns if present. Next, with a sharp knife, gently scrape away a 2-inch section of green on the stem. Scrape until the white is exposed. Then pierce the bottom of the stem with the knife. Where the stem has been scraped and pierced, this is where roots will form. After the stem has been scraped and pierced, dip or coat the entire exposed and scraped section of the stem in honey.

Many gardeners will use hormone rooting powder instead of honey. The hormone powder crowd insists that it's superior. But I've never noticed any difference. If anything, I think honey works better. Honey has antiseptic properties and gives the exposed regenerative plant cells a little extra sugar. Honey also clings to the stem cuttings better than rooting powder.

Once the stem cutting has been prepared, it is next inserted into wet, sterile potting soil. Allow about 1 inch to 4 inches of the cutting to remain above the soil line. Be sure to tamp the soil down firmly around the stem cutting to remove any air pockets. After the cutting has been planted, the 2-quart Wide Mouth Mason jar is placed over the cutting. Be sure to leave plenty of headroom. The pot is next flooded with water and moved to a semi-shaded location for about 8 to 12 weeks.

The soil around the cuttings must stay moist but not overly wet or saturated. Don't let the cuttings dry out, or they may die. On average, it takes about 3 to 9 weeks for cuttings to begin to grow roots. Some plants will take much longer. Bay laurel plants can take up to one year to begin to form a viable root system. This is where it's essential to understand the habits of the plant you're attempting to propagate. Usually, I start to check for root growth about week 4 or 5. Once root growth has commenced, remove the Mason jar for a few hours during the day so that the baby plant can grow accustomed to normal air circulation. If the newly developed plant starts to

66

The Complete Granny Miller

look stressed or wilted at any time, replace the Mason jar for another week. If any flower buds should begin to form on the stem cutting, remove them at once so that the young plant conserves energy and puts all its effort into growing leaves and roots.

Aphids

What They Are

Aphids, sometimes called plant lice, are tiny sap sucking insects. There are many different species of aphids, and they come in several sizes. Aphids may be green, red, brown, or black, and sometimes a fuzzy white, pink, or gray. All aphids, no matter how small, have tiny pear-shaped bodies with long legs and antennae. Aphids commonly begin to infest plants during late spring. This is especially true when plants have been side-dressed with composted manure or fertilized heavy. High levels of nitrogen encourage aphid reproduction. Aphids tend to do the most damage when temperatures are between 65°F to 80°F.

During the hot summer months, apple trees can suffer from wooly apple aphids, green apple aphids, and rosy apple aphids. Ants and aphids are often partners in crime. So, if you see ants in large numbers crawling up a vegetable plant, tree, or bush, you can be pretty sure that aphids are present. In some situations, ants will watch over and protect aphids because ants like to feed on the 'honeydew' that aphids excrete. Aphids will not kill a large plant, rose bush, or fruit tree. But a heavy aphid infestation can cause the leaves of a plant to curl, deform flowers, and weaken plants. Aphids spread plant viruses. Young plant seedlings are especially susceptible to aphid damage. Cabbage, beets, melons, squashes, pumpkins, potatoes, and beans often suffer from aphid-transmitted viruses.

Ways to Control Them

Aphids are easy to control. It seems that every gardener has their favorite aphid remedy. For a few plants with a light infestation, aphids can be removed by hand or sprayed off with a stream of water. If you are going to control aphids with a spray of water, it's best to do it early in the day so that the plants can dry in the sun. Sun drying reduces the possibility of fungal diseases. For more serious infestations, especially in fruit trees, aphids can be controlled by means of chemical control, like Malathion, permethrin, and acephate.

For gardeners who prefer a more natural approach to pest control, soapy water delivered by means of a pressure tank sprayer or a small squirt bottle works well. For fruit trees, I use a ¼ cup of Dawn dish detergent in a 3-gallon spray tank of water. Spray only the trunk of the trees and never the leaves or fruit. Take care and don't spray soapy water on stressed plants or when temperatures are above 90° F. No matter what type of spray you choose, be sure to spray the underside of the plant leaves where aphids tend to hide. Aphids do have natural enemies and predators. Sometimes, biological control can be effective in reducing aphid populations, especially in small backyards. Ladybugs, lace wings, and syrphid flies are the best-

The Complete Granny Miller

known predators of aphids. Ladybugs can be ordered by mail, and when properly handled, can afford some relief to a heavy aphid infestation.

How to Plant a Bedding Plant

I learned how to plant bedding plants from a Greek immigrant with an Old-World green thumb more than 40 years ago. I have taught his method to everyone I know who asks. It's really pretty simple.

1. First, dig a hole large enough to accommodate the root ball of the plant.
2. Next, fill the hole with water and set the plant inside the hole to within 1 inch of the bottom leaves.
3. Next, hill up the soil around the plant and pack it down firmly.

This method works well because the water in the hole acts to pull the roots downward into the earth and provides vital moisture. Transplanting can be a shock to young plants, but the extra water will help them recover. When planting bedding plants, try to pick a cloudy day, or in the evening, or right before it's going to rain. Never plant during the heat of the day, and keep newly transplanted plants well-watered.

HERBS

A herb garden has always been an essential part of my gardening life. I grow vegetables because I need to eat. I grow flowers because I love fresh-cut flowers in my home. But I grow herbs - well, just because!

When I was younger, I kept an enormous herb garden that divided the various herbs by their uses and purposes. I grew medicinal herbs, herbs for natural dyeing, herbs for scent, and culinary herbs. Now that I'm older, I don't have the time or the energy to keep up with such an ambitious garden scheme. Today, I only keep one small kitchen herb garden to supply my needs. Situated right off my kitchen, it's a convenient location when I'm cooking. If I need fresh parsley, basil, sage, rosemary, mint, or any other herb, it's just a quick trip out the door. In my herb garden, there is a path made of flat stones to help keep my feet dry and clean while I gather herbs or work in the garden.

In general, herb gardens that back up to a house or any other building do best when facing south. That's because a southern exposure usually ensures full sun and keeps the herbs out of the shade or shadows. Sunshine keeps herbs healthy and vigorous. Most common garden herbs need about 8 to 12

The Complete Granny Miller

hours of full sun per day. The exceptions that I can think of are the various types of mints, lemon balm, and sometimes sweet woodruff. Those herbs will do quite well in a moist, semi-shaded location. The majority of kitchen or culinary herbs require average to poor soil.

Borage

Borage is an annual herb with coarse, hairy leaves and stems. It grows 2 to 3 ft. tall and often spreads as wide. Borage (Borago officinalis) is also known as 'starflower' and is sometimes called 'bee bread'. That's because of the sky-blue star-shaped flowers and because borage is a bee magnet. The honey made from borage flowers has a reputation for being very fine and tasty. Many people plant borage in the vegetable garden for pollination purposes. I grow borage in my herb garden mainly for the beautiful flowers. I like the color blue. And if you ask me, you can't get enough good blues in either the flower or herb garden.

Borage is considered to be a 'potherb'. If you don't already know, a potherb is a plant whose leaves, stems, or flowers are cooked and eaten, or used fresh to season food. Fresh mint is considered to be a potherb. Kale, spinach, beet greens, and many other leafy plants are examples of potherbs that are usually cooked.

At one time, fresh borage flowers were preserved by being candied. During the Middle Ages, wine was infused with borage to banish gloominess and melancholy. Back in the 1960s, it was popular to freeze the star-shaped blue flowers into ice cubes to make pretty summertime drinks. Borage flowers have a slight cucumber-like taste. Small amounts can be used in summer salads to impart a little zest and zing, as well as beautiful color.

In recent years, there has been growing interest in borage seed oil. Borage seed oil contains gamma-linolenic acid (GLA) and appears to have a mild anti-inflammatory effect. Borage seed oil is soothing to eczema, seborrheic dermatitis, and other skin disorders. It is a source of prostaglandins used to treat menstrual disorders.

Borage is easy to grow and prefers to be directly sown in the garden. It thrives in a dry and sunny location. Borage readily self-sows and comes up every year from the seeds that were dropped from the previous year. Flowers are picked as they open, and leaves are harvested fresh.

Chamomile

Chamomile is a common bushy garden herb with small daisy-type flowers. It has a fragrance similar to apples. It is easy to grow and needs full direct sun and average soil conditions. Chamomile will readily volunteer new plants every year and can be treated like a perennial. There are a few different types of chamomile. The type that I have growing in my garden is ordinary German chamomile.

Chamomile is easily started in the garden by simply sprinkling the seed

The Complete Granny Miller

onto prepared ground and gently raking it into the soil. It is okay if the seed isn't completely covered or if it's covered lightly. Just tamp the seed firmly so that it makes good soil contact and water well. Before too long, you'll have more chamomile than you'll know what to do with. But be careful. After a few years, chamomile can take over your garden or lawn if you don't weed out some of the plants regularly. The yellow and white daisy-like flowers of chamomile are usually dried and infused in water to make a tea. The tea is most often sweetened with honey or sometimes lemon and cane sugar. Chamomile has been used for centuries to calm fretful and colicky babies, and is a well-known and safe sleep aid for adults. It can be helpful in soothing heartburn, nausea, and the miserable after effects of vomiting. Some people will use chamomile as a gentle wash for hemorrhoids. The soothing properties are sometimes used to treat minor skin irritations. Chamomile is also used as a mouthwash or hair rinse for blondes.

To collect chamomile flowers for drying, pick a hot, sunny, dry day. Make sure the morning dew is off the plants. You can pick the flowers individually, or you can use a rat-tail comb with a double row of teeth and comb the flower heads off the plant. Once collected, chamomile flowers can be dried on a screen, in a paper bag, in a lined food dehydrator, or in a slightly warm oven. Take care, as chamomile dries quickly, and it is easy to over-dry the flower heads. When the flowers are completely dried, store them in a tightly covered container in a dark, dry location.

Rosemary

Years ago, when I first moved to Pennsylvania from Tennessee, I was disheartened to find that some herbs that I had always grown as perennials would not stand the harsh winters of Western Pennsylvania. Here in agricultural zone 5b, the one perennial herb that I have to replant every year is common rosemary.

Rosemary is a wonderfully fragrant herb with small, pointed, evergreen leaves and is native to the Mediterranean basin. In Western culture, it is often associated with the Virgin Mary because of its blue flowers.

Rosemary is used in cooking to season lamb, pork, chicken, and certain stews. But be careful when you cook and season with it. A little rosemary goes a long way. It's easy to overpower a recipe with it.

The essential oil of rosemary is used to make Hungary Water, incense, and perfume. Rosemary has a traditional use for restoring memory, curing headaches, and, not so long ago, it was burned in sick rooms with juniper berries as a disinfectant. Rosemary is associated with love, steadfastness, and faithfulness. At one time, it was commonly woven into bridal wreaths. In some parts of Wales, it is still the custom to give small sprigs of rosemary to funeral guests and mourners to throw into a freshly dug grave.

New rosemary plants are best started from stem cuttings and not from seed. Rosemary that is grown from seed is often inferior to that which is started from a slip or a cutting. Rosemary is quite happy to live as a potted plant and often can be moved indoors to a sunny windowsill during the winter.

The Complete Granny Miller

Like most herbs, rosemary requires a full sun location for the best fragrance. Ordinary to poor soil and not too much water will keep rosemary healthy and vigorous.

Sweet Woodruff

Sweet woodruff is a low-growing perennial herb that thrives in a moist and semi-shaded location. It flowers in May and is sometimes known as sweet scented bedstraw. The leaves and small white flowers are often used in sachets and potpourri. The sachets are used in drawers and linen closets to give a fresh scent to clothes and to keep moths and other insects away. When first gathered for drying, the leaves and flowers of sweet woodruff have little odor. But once the leaves and flowers begin to wilt and are properly dried, the scent becomes more pronounced and can last for years. Sweet woodruff is sometimes used to flavor ice cream and is the herb that gives May wine its distinctive flavor.

Basil

There are many different varieties of basil. But probably the most popular form of this herb is sweet basil. Sweet basil is the herb for all things tomato. Basil is an easily grown annual plant that is highly susceptible to cold. In fact, planting basil too early in the spring is the leading cause of failure when growing this herb. Cold air and cool soil will stunt the growth of basil. Basil plants that have been exposed to cold temperatures never seem to recover from the setback fully.

If you want to grow basil in your herb garden, wait until the spring weather has settled. Basil seed usually germinates in 5 to 7 days when the soil temperature is above 68°F. Basil leaves can be harvested for fresh use with tomatoes, or the leaves can be dried for winter storage. Basil needs full sun and benefits from an occasional watering.

Mint

Like basil, there are numerous different varieties of mint. One characteristic all mints have in common is a square stem. Most mints are highly aggressive in their growth habit. It's a good idea to give mint its own space, separate from other plants. Mints grow and spread by sending out tenacious runners and roots. Mint will tend to escape from where it was initially planted. It never fails to grow where you don't want it to.

All mints grow best in full sun, but they can tolerate a little shade. Unlike many kitchen herbs, mints do best with rich soil and more water. To harvest mint, simply snip the leaves. Most mints are easy to dry. Peppermint makes a tea that is soothing to upset stomachs and can help break a fever.

The Complete Granny Miller

Catnip

Catnip or catswort is a member of the mint family. Like all mints, catnip grows best in full direct sun and will tend to take over wherever it is planted. Catnip is a drought-tolerant, hardy perennial with slightly heart-shaped, saw-toothed leaves. Catnip contains the substance nepetalactone, which is an attractant for most felines - domestic and otherwise.

Catnip excites cats but tends to have a calming effect upon the human digestive system. It is a well-known folk remedy for colicky babies.

Parsley

In my area of the world, parsley is usually grown as a biennial herb. If you don't know, a biennial plant is a plant that takes two years to complete its biological life. When parsley is planted from seed, the first year of its growth is used to establish a tap root and leaves. During the second year of life, a parsley plant grows larger and will go to seed.

The two main types of parsley that are usually offered through seed catalogues are curly leaf parsley and flat leaf parsley. I routinely use both types of parsley in my kitchen.

Many cooks have a preference for flat-leaf parsley in cooking and reserve the curly-leaf type for food garnish. Whatever type of parsley you prefer, be aware that parsley has a reputation for being slow to start from seed. In fact, the old saying about slow-starting parsley is,

"Parsley seed goes nine times to the devil."

I'm not sure about that. But what I am sure of is that parsley takes about 2 ½ to 3 weeks from the time it is sown to germination and appearance above the soil. Like most kitchen herbs, parsley needs full direct sun and average soil.

Oregano

Oregano is a member of the mint family. It is the herb that makes pizza taste like pizza. There are several different varieties of oregano. But the one type that is most often grown as a common culinary herb is Greek oregano. Greek oregano is a low-growing perennial herb that tolerates arid conditions. Oregano produces a small pink flower that is attractive to bees. The leaves of oregano are easily dried and stored for winter use. The flavor of oregano leaves is enhanced by drying.

Common Sage

Common sage is the herb that many people associate with turkey stuffing and breakfast sausage. Sage is a powerful culinary herb, and a little bit goes a long way. Be careful when cooking with sage. It's easy to overdo it and ruin the dish.

Sage is most often grown as a perennial herb and it is easy to start from seed. Sage seeds are large and round and tend to germinate within 7 to 15 days. Sage grows about 24 inches tall and produces grayish-green leaves that are oblong and slightly pointed. It blooms with a lovely pinkish-purple flower that is attractive to bees. Common sage will usually survive the harsh winters here in Western Pennsylvania. In early spring, cut sage plants to the ground to rejuvenate them. They will grow back up beautiful and bushy. Like most kitchen herbs, common sage prefers full sun and average garden soil.

Drying Herbs

If herbs are going to be dried for storage, they need to be perfectly dry when you harvest them. Pick them on a sunny day if possible, and after all the morning dew has dried. Only harvest the best quality herbs. Drying won't improve the quality of an unhealthy plant or an herb that is past its prime.

When to Dry

All herbs dry best when they are picked during a waning moon and in the signs of Aries, Leo, or Sagittarius. In fact, the further away you get from the full moon, the better they will dry. Herbs that are harvested when the moon is waxing or full, or harvested when the moon is in the sign of Cancer, will have a tendency to go moldy and to spoil. To dry herbs for storage, wait for the full moon to pass and begin to wane. You can consult any good almanac for information about daily lunar cycles and the moon's age and location.

How to Dry Herbs

I use a cheap plastic electric dehydrator for most kitchen herbs. Parsley, lemon balm, bay leaves, rosemary, sage, tarragon, sweet basil, and all kinds of mint are dried to perfection in a food dehydrator. Depending upon the relative humidity, most herbs will dry within 2 to 24 hours. Herbs are dry when the stems break easily and the leaves or flowers readily crumble. A barely warm electric or gas oven also works well for drying.

With small leaf herbs like marjoram, thyme,or oregano, I dry them on an old window screen. Screen drying seems to work better than a food dehydrator for small-leafed herbs. When plant material dries, it shrinks. Small-leafed herbs, once they shrink, tend to fall through a plastic dehydrator tray. For screen drying, I use ordinary window screens laid across two sawhorses. The screens are placed in a warm, sheltered location until the herbs are completely dried.

Long-stemmed or large-leafed herbs like dill, mint, sweet basil, fennel, anise, bee balm, and lavender are easily dried by being tied into bundles and hung upside down in a warm location. I use new rubber bands or garden

The Complete Granny Miller

twine to hold the bundles together.

With herbs that have small seed heads like dill, anise, or coriander, I place a small paper bag over the heads and tie the bag securely to the stems of the herb with a string. That way, when the seed heads begin to dry and fall, the seeds that fall will fall into the bag and not onto the floor to be wasted.

Glass Mason jars are perfect for storing herbs. But plain paper bags work well, too. Dried herbs are stronger-tasting than fresh herbs. Be sure to consider that when you cook with them.

Medicinal Herbs

If herbs are to be used fresh or for medicinal uses, they are most flavorful and potent when harvested while the moon is waxing or full. Folk wisdom informs us that herbs to be used fresh for healing purposes are best picked in the spring or early summer during the light of a full moon, or when the moon is in the sign of Scorpio.

The Complete Granny Miller

Chapter 4

"Small opportunities are often the beginning of great enterprises." Demosthenes.

The Homestead Orchard

Fruit Trees

Fruit trees are a wise investment for most homesteads. In fact, fruit trees were some of the first improvements my husband and I made when we began to remake the family farm over 30 years ago.

Back in those days, the local 4-H sold fruit trees as a club project for $7 a piece. Each year, I could only afford to buy 5 or 6 trees at a time. Money was tight. My husband and I had just started to remake his old family homestead. The farm had been left idle for an entire generation, and we had more repairs and bills than we had money for. Every spring, I scrimped on groceries to pay for the fruit trees and would drive on snow-covered back roads to the next county to pick them up. It was some trouble - but it was worth it.

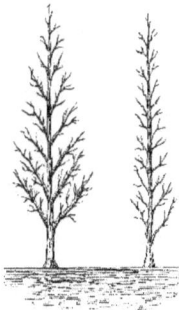

As a new homesteader, I believed then, as I do now, that it is necessary to plant fruit trees first when settling in at a new place. I took to heart a lesson from the early settlers of Pennsylvania. They depended upon apples for fruit, cider, and vinegar for food preservation. For them, apple trees came before anything else and were planted first on a new homestead. Fruit trees are an important part of the foundation upon which home food production and self-reliance are built.

Apple Trees

Apple trees can last a lifetime. And depending upon the variety, they'll take between 2 to 10 years to mature and to produce fruit. If you're considering adding a few apple trees to your backyard garden but are hesitant, I hope the following will encourage and reassure you that a small home orchard is a worthwhile investment. Apple trees are much less trouble than most people imagine them to be.

Location

When planning a home orchard, picking the correct location is essential to its success. Unlike a vegetable or flower garden, an orchard cannot be

The Complete Granny Miller

moved to another location once it is planted. All fruit trees need at least 8 to 10 hours of full sun every day. There should be plenty of space between the trees so that the air can freely circulate around them. Never locate apple trees or any type of fruit tree in a low-lying wet area. You also want to avoid areas where frost can run down a hillside or collect in pockets.

Size of Trees

It's vital to pick the right size apple tree for your location. With modern apple trees, there are basically three sizes: dwarf, semi-dwarf, and standard. Dwarf apple trees are the smallest size, and standard apple trees are the largest. Most commercial apple orchards are planted in semi-dwarf trees.

Modern Fruit Trees Have Two Parts

Modern apples are actually grafts that are made in two parts: the scion and the rootstock. The two parts together make one tree. The scion and not the rootstock determine the variety or type of apple. The scion and the rootstock are two completely different things.

The scion section is the top part of the tree. It's the part that's completely above ground and is made up of the leaves, limbs, and branches. The scion bears the apples or other fruit. Scions are grafted onto different types of rootstock. With modern apples, it's the type of rootstock that determines the size of the tree.

Rootstock is what grows beneath the ground, and you cannot see it. Different types of rootstock control how large a tree will eventually become. It's the rootstock that determines whether or not a tree is dwarf, semi-dwarf, or standard. If you look carefully at the base of a fruit tree, you can sometimes see where the tree was grafted. The scion union graft in apples is a slight bump area about 1½ to 4 inches above the roots. Many different types of apple varieties are available on dwarf, semi-dwarf, or standard-size trees. Since rootstock determines the size of the tree, you can have almost any type of apple you like. The size of the tree does not determine the Apple variety.

When you will get your first apples depends on the variety of apple and the size of the tree. Just remember that the bigger the tree, the longer the wait. A standard-sized apple tree can take up to 7 years to produce apples. A semi-dwarf tree usually produces its first fruit within 2 or 3 years. A dwarf apple tree can produce fruit the first year after it's planted. Dwarf trees are popular for that reason. Dwarf trees are perfect for small areas where space is a problem. They can do well on decks and patios when planted in large containers. And just so you know, some people say it takes at least twenty leaves on a tree to produce one apple. So, according to that theory, you'll need at least one hundred leaves on your apple tree to get five apples.

Pick the Right Variety

There's a big difference in apple varieties. And different apples are used for different purposes. There are cooking apples, eating apples, and apples for storage. Certain types of apples make better cider and vinegar. The variety of apple tree you choose to plant depends on your intended purpose. Apples don't all ripen at the same time; nor do they have the exact cultivation requirements, disease resistance, or spurring habits. When selecting an apple variety for your garden farm, keep in mind what types of fungus or diseases are prevalent in your area. A trip to a local apple orchard or grower can be helpful in determining this information. Here's an example of why local information is critical.

Let's say that apple scab is a problem in your area. For that, you would do better to pick a scab-resistant apple like Liberty or Red Free. If fire blight is a problem in your location (it's a problem where I live), Prima would be a good choice, and you might want to forget about Rome apples. Only by knowing your specific piece of ground can you make an informed decision about what particular variety of apple is best for your place and your location.

When buying fruit trees, I strongly caution you to avoid potted trees from big box stores like Walmart, Lowe's, or Home Depot. Don't waste your money. Only buy bare-root fruit trees from a reputable nursery. Adams County Nursery and Stark Brothers are both good companies to do business with. If you do decide to buy an already potted fruit tree from a local nursery, just make sure that it is guaranteed. Most reputable companies will guarantee their fruit trees for at least one year.

It's my opinion that the best trees for the home orchard are 1-year-old whips or 2-year-old bare rootstock. A small tree with a sound root system will always transplant better than a larger tree. Only plant fruit trees in the spring or in the autumn when the trees have no leaves on them. Planting any type of deciduous tree while it's in leaf is a risky business. When planting any type of tree, ensure it is planted in a large enough hole. You don't want to crowd or jam the roots into a too-small hole. Try not to break the roots. There's an old expression about planting trees:

"You'll do better to plant a dime-size tree in a dollar hole than to plant a dollar-size tree in a dime hole."

How and When to Plant an Apple Tree

Take my advice and consult an almanac before you break out the garden rake or shovel. That's because the best time to plant an apple tree is in the moon's 3rd quarter. All trees and perennial plants will form a deeper and sturdier root system when planted during the decrease of the moon. If at all possible, plant trees in the sign of Taurus and avoid the fruitful or watery signs such as Cancer, Pisces, or Scorpio. Trees planted in the fruitful signs tend to make vigorous top growth at the expense of the roots.

When planting young trees, the formation of a strong root system is vital to their future. Autumn is my preferred planting time for all trees. But spring planting can be just as successful. The advantage of a fall planting is that young trees have an extra two or three months to develop strong roots before the warm season arrives again. Fruit trees benefit from a longer season of cool, rainy weather after planting. Most mail-order nurseries are set up for a spring rush. But sometimes they do offer fruit trees for sale in the latter part of the year, but the selection is not as good.

When you get your apple trees, it is imperative that they do not dry out

The Complete Granny Miller

before planting. Keep them in a dark, cool location and well wrapped until planting. Wait to plant them until the soil is open and workable. Try to pick a cool overcast day in the moon's 3rd quarter as suggested. Avoid planting too early in the spring or when the ground is saturated and wet. This is especially true for heavy clay type soils.

Make a Hole Six Inches Wider Than the Roots

To plant an apple tree correctly, you'll first need to dig a large and deep enough hole. The hole should be at least 6 inches wider than the broadest part of the roots. If your tree is bare-root, it's a good idea to make a small saddle of dirt for the roots. A dirt saddle is a small hill or hump inside a planting hole. To make a dirt saddle, reach inside the hole and form a small conical-shaped dirt hump. To plant the tree, place the center of the root mass on top of the dirt saddle. Allow the roots to cascade down over the hump. If your tree is potted, you will need to carefully remove it from its pot before it is set into the hole. If the tree is wrapped in burlap or potted in a fiber pot, you don't have to do anything. Just set the pot in the hole - burlap and all.

When you have the tree in the hole, try to place the sturdiest part of the scion graft so that it faces the prevailing wind. When I first planted my orchard many years ago, I really didn't know what I was doing. I planted some of my trees with the weakest part of the scion graft facing into the prevailing wind. Over time, some of the trees began lean because the wind was blowing against the graft.

Once the tree is placed into the hole and the scion union is facing the right direction, you'll want to flood the hole with water. I usually pour an entire 5-gallon bucket of water into the hole. Pouring water into the planting hole serves two purposes. The first is that it helps to pull all of the roots downward and prevents air pockets. Air pockets can sometimes occur when the soil is placed back into the hole. Secondly, the water will prevent the tree from drying out in the event of an unseasonably warm spell and reduces the stress of transplanting. Allow some of the water to recede in the hole before raking or shoveling the dirt back into the hole. It's okay to place the grass or turf back in the hole. Just turn the turf upside down and stuff it back in. Gently but firmly tamp the dirt with your foot around the entire trunk of the newly planted tree. You want to ensure that there are no dead spaces or air pockets in the hole.

After Planting, Prune the Tree

Here comes the hard part. After your tree is planted, it needs to be immediately pruned, keeping a 3 1 ratio. That means keeping the top leader or the branches one-third the length of the roots. When planting a 1-year-old un-branched whip, cut the top of the tree off at 30 to 36 inches above the ground. I know you won't want to cut the tree. But just do it.

Remember, any pruning done in late winter or early spring encourages branching. By cutting the tree off above the ground in the spring or late

The Complete Granny Miller

winter, you are encouraging the tree to begin to branch out and form its scaffolding system. The scaffolding system is the design of the limb structure that will grow to become a permanent part of the tree. Any pruning done in the summer or early autumn discourages growth. This is good information to know, as it is key in case water sprouts become a problem. Water sprouts are long branches that sometimes grow up from the base or trunk of a fruit tree.

Water and Spray

After you have planted the apple tree, you'll want to keep it well watered at regular intervals during its first spring and summer. You also may need to spray for summertime pests. I try to spray my trees once a week during the summer with soapy water. I use a 3-gallon spray tank mixed with water and ¼ cup of Dawn dish detergent. Spray only the trunk of the trees and never the leaves or apples. The trunk of the tree is like a bridge for bugs. By protecting the tree trunk, you are protecting the tree against pests. Protect the trunk of an apple tree and you protect the entire tree.

How to Feed and Fertilize Fruit Trees

It's essential to feed fruit trees every year. But feed them only once a year. Fruit trees should never be over-fertilized. The best time for feeding fruit trees is when they are at 'silver tip'.

Silver tip is when the buds are swollen and look silvery steel gray in color. I use calcium nitrate 15.5-0- 0, sometimes called Norgessalpete or Norwegian saltpeter, to feed the trees in my orchard. The trees are fed by sprinkling the fertilizer from the base of the tree to the 'drip line'.

The drip line for a tree is directly underneath the farthest extending branches.

Fruit trees are fertilized with calcium nitrate at the following rates:

- 8 lbs. for large trees
- 6 lbs. for medium trees
- 1 lb. for small trees

Apple Trees & Expert Advice

When I was younger, I wasn't as smart as I am today. Back in those days, I believed just about everything the 'experts' from major universities and the local agricultural extension office told me. Too bad for me.

Because when I planted my small commercial apple orchard, I didn't have any practical experience with growing apple trees. I was pretty much a babe in the woods. I relied upon books and pamphlets, and the advice from the local agriculture extension office and Penn State pomologists.

Back then, the expert advice for spacing semi-dwarf apple trees was 12 ft.

The Complete Granny Miller

to 15 ft. apart on center. The experts at Cornell University and Penn State replaced the good advice my father-in-law had tried to give me while I was planting trees.

My father-in-law, who never grew an apple tree but had been a lifelong gardener, came upon me one day while I was planting a bunch of 1-year-old apple whips. My father-in-law advised me that the trees were being planted too close together. He suggested that they be placed farther apart so as not to crowd each other once they attained full size. I dismissed his suggestion due to his lack of formal education and practical experience in fruit orchards. I explained to him that 15 ft. spacing was what the extension office pamphlet recommended, and that's what I was going to do.

He had no use for newfangled Penn State notions and tried to convince me to add more space. He said that he may not have ever read a book on orchard management, but he knew something about the way trees grow. I was determined to do it my way and would not entertain his suggestion. He wisely shrugged his shoulders and walked away, leaving me with my expert advice and a future problem. Time has proved him right and the experts wrong.

Life experience has since taught me that the correct spacing for semi-dwarf apple trees is a minimum of 18 ft. apart on center, with 22 ft. being ideal. Because my apple trees were planted so close together, they are hard to manage properly. Apple production has been on a steady decline for the last five years or so. I could have been spared the trouble and heartache of destroying mature apple trees if only I had taken my father-in-law's advice. Gene Logsdon wrote something years later on this very subject. It was too late to be of benefit to me. But I thought Gene's advice may be of benefit to you, so I'll share it.

'My wife and I produce most of our food, and some for our children's families, using knowledge we gained from our parents.

Not a one of our forebears ever cracked an agronomic textbook or knew the Latin name of a single plant. My father and mother and both grandfathers and grandmothers and my father-in-law and mother-in-law all held agricultural advisers in distain.

Tradition supplemented by our own experience and that of other gardeners and farmers, has been the key to our food-growing success. Thousands of books by gardeners and farmers pass this knowledge on to anyone who wants it. To this day, after forty years of avidly reading and searching the realms of "modern" agriculture for information, I have found little knowledge beyond oral tradition that helps us produce food any better. And a whole lot that encourages us to produce it worser.

The keys to agricultural success, apart from common sense, were articulated by Virgil, and he got them from the Greeks, who in turn got them from the Orient, where forty centuries ago China supported a population far denser than ours today, with gardens.'

Fruit Tree Grafting

Fruit tree grafting is an easy-to-learn homestead skill. As previously mentioned, most modern apples, grapes, and other fruit trees are grown from two parts: a rootstock and a scion. When the scion and rootstock are joined together, they form a new plant or tree. Grafting is the mechanical process by which the rootstock and a scion are joined together to form a new plant. Apple or other fruit scions can be free for the asking in the springtime. Rootstock can be ordered by mail or online for about $3 a piece (2016 price).

There are a few different methods of grafting. Each method has an advantage and a disadvantage. The method that I think is the easiest for the beginner to learn and to master is the 'cleft graft'. With a cleft graft, a scion and root stockof similar diameter are joined by way of a wedge cut that matches the cambium layers of both the scion and the root stock.

The graft is then secured with wax, tape, or by other means for a few weeks, while the tissues grow to form a union. Cleft grafting can be easily done with a sharp knife. But I have found that beginners have the most success with grafting shears. Grafting shears are a hand tool that helps to make tight-fitting and interlocking wedge cuts that fit together like a puzzle piece.

How to Graft a Fruit Tree

You will need:

- Scions
- Rootstock
- Grafting wax, electrical tape, or duct tape
- Grafting tool/shears

In the early spring, choose a scion that has at least 3 or 4 well-spaced, plump buds. Choose the variety of rootstock that you desire for your location. Cut the rootstock with the grafting tool so that it's about 12 to 18 inches long. Flip the grafting tool around and cut the scion to the same diameter as the rootstock. The two pieces will now fit together like a puzzle.

Join the two pieces and secure with wax or tape. Plant the new trees in individual pots or the ground. Keep them well watered for at least 4 to 6 weeks or until new leafy growth is noticed above the graft area.

New growth above the graft area indicates that the graft is successful. Growth below the scion with no top growth indicates that the graft has failed.

Grafts are most successful when performed during the moon's 1st and 2nd quarters and in the sign of Cancer. Any good current almanac will have that information. Cleft grafting is an affordable way to produce lots of trees, vines, or other plants.

The Complete Granny Miller

Mummy Fruit

Mummy fruits are brown, shriveled fruits that don't drop from a tree during the autumn. They contribute to disease and fungus cycles in fruit trees. Fire blight and bitter rot are two important orchard diseases that can spread through mummified apples and other fruit. Wet or humid weather tends to promote both conditions. Troubles are made worse by not removing mummified fruit from trees.

Mummified fruit should be removed from all trees before spring so that disease does not spread in the orchard. When mummy fruit is removed, it should be burnt well away from other fruit trees. That's because the smoke from burning fire blight limbs or mummy fruit can re-infect the orchard. Good sanitation is an integral part of orchard management and is a critical component in the reduction of bitter rot, fire blight, and other diseases.

How & When to Pick Pears

Pears, unlike apples, are best picked while they are still slightly immature. The finest quality pears for fresh eating or home canning are those that are not fully ripened on the tree. A pear that is allowed to ripen on a tree often has a mealy texture and a soft or mushy core. That's because pears tend to ripen from the inside out. Often, when a pear looks soft, ripe, and ready on the tree, the interior is usually on its way to rotten. A pear is ready to be picked when it snaps away from the tree while being lifted towards the sky. To ripen fresh-picked pears, place them in a cool, dark location like a cellar. If you don't have a root cellar and only need to ripen a few pears, place the pears in a brown paper bag with a ripe apple or banana. The ripe apple or banana gives off ethylene gas, which will stimulate the ripening of the pears. Pears are ready for canning and for fresh eating when the flesh around the stem area gives slightly under firm pressure.

Forced Branches

I try to have fresh flowers or a little bit of nature indoors during every season of the year. During the winter, my favorite thing to bring indoors is the branches of flowering trees and shrubs that lie dormant waiting for spring. The flower buds on spring-blooming trees and shrubs are fully and perfectly formed by the middle to end of autumn. All that's required for the branches to bloom and flower is warm temperatures and water.

Usually, it takes 6 to 8 weeks of cold weather before the flower buds on most trees and shrubs awaken from their dormancy. The branches have a built-in inner clock. The inner clock is nature's way of protecting flower buds from opening prematurely during a warm spell. If flower buds open too soon, they are destroyed by the returning cold weather. It's one of nature's many ways to ensure that a tree or shrub will set fruit or go to seed. Here in Western Pennsylvania, I can bring flowering branches into the house for forcing any time after the first week of January. With proper

The Complete Granny Miller

preparation, they will bloom.When I go scouting for branches, I always carry sharp, heavy-duty pruners with me. Frozen branches can be quite hard to cut through.

I look for well-formed, large buds and try to pick branches that will fit into the particular vase that I have in mind. I never cut too many branches from any given tree or shrub so as not to reduce the summer or fall fruit harvest. A little cutting goes a long way.

After the branches enter the house, the bottom of the stem requires special treatment. Any cut made on a branch will seal over. So, it's important to either scrape away some of the skin on the branch or to smash the end of the branch before it is placed into water. By opening up the bark, water can travel up the branch, keeping it fresh and alive. The buds of certain types of branches require a bit more care to ensure a nice bloom. Apple, peach, plum, pears, and forsythia should be soaked overnight in room temperature water before they are placed into a vase. A bathtub or wallpaper tray is perfect for this. Once arranged in a vase, the branches should be kept in a cool location and away from direct sunlight. We are trying to mimic spring. Spring comes slowly, and it doesn't come all in one day.

When the flower buds begin to fatten or plump up, they can be moved to filtered sunlight. In fact, a modest amount of sunlight will be necessary for the flower buds at this stage. A little sunlight helps them to make a good color.

The Complete Granny Miller

Chapter 5

"It is always the simple that produces the marvelous."- Amelia Barr.

Something Good To Eat

Gardening farming, by its nature, often results in an excess of food that must be stored or preserved in some way for later consumption. Home canning has had a revival in the last ten years or so. So, it seems to me like home canning might be a good place to begin this chapter.

A Canning Primer

The Two Types of Home Canning
For home canning, foods are divided into two distinct categories:
- High-Acid Food
- Low-Acid Food

The two types of foods are processed differently.
- High-acid foods are those foods that have naturally occurring acids. Most fruits and the majority of tomatoes are considered to be high-acid foods. Foods such as pickles, where vinegar or lemon juice has been added, are considered to be high-acid foods. Many fermented foods, such as sauerkraut, are also high-acid foods

- Low-acid foods are those foods that have little naturally occurring acid. All common vegetables, meat, fish, poultry, mushrooms, and milk are consideredto be low-acid foods.
  The only dependably safe method to home can low-acid foods is with a laboratory tested recipe and with a pressure canner in good working order.

It's the level of acidity in food that determines whether or not certain harmful organisms can survive and reproduce in a sealed canning jar. Potentially harmful organisms in **high-acid** foods **CAN** be dependably destroyed at 212° F in a boiling water bath canner.
Potentially harmful organisms in **low-acid** foods **CANNOT** be destroyed at 212°F in a boiling water bath.

When processing low-acid foods, the temperature inside the canner **MUST** reach at least 240°F. Heat must go into the inner core of the food packed into the canning jar. The temperature also must be maintained for the proper amount of time for the particular food being canned.

At sea level, water boils at 212°F. Boiling water cannot guarantee the complete elimination of harmful bacteria or spores that may be present in low-acid foods. It is steam under pressure that drives the temperature up inside a pressure canner to 240°F - a temperature well above the point of boiling water.

Make Adjustments for High Altitudes

Water boils at a different temperature depending on the altitude. Most laboratory tested recipes for home canning are written for locations at 1000 ft. sea level or lower. At sea level, water boils at 212F°.

But depending upon the altitude, water will boil at a lower temperature. For boiling water bath canning, the processing time must be increased at high altitudes.

Altitude and Elevation Adjustments Needed for Water Bath Canner	
0 - 1000 ft.	No adjustment
1001 - 3000 ft.	Add 5 minutes
3001 - 6000 ft.	Add 10 minutes
6000 - 8000 ft.	Add 15 minutes
8001 ft. - 10,000 ft.	Add 20 minutes

For pressure canning at high altitudes, the pressure inside the canner must be increased. But the processing time remains the same.

Altitude and Elevation Adjustments Needed for a Pressure Canner		
Altitude in Feet	Dial Gauge	Weighted Gauge
0 – 1000	10 psi	10
1001 – 2000 ft.	11 psi	15
2001 – 4000 ft.	12 psi	15
4001 – 6000 ft.	13 psi	15
6001 – 8000 ft.	14 psi	15
8001 - 10,000 ft.	15 psi	15

The Complete Granny Miller

Canning FAQ

Is It a Canning Jar or a Mason Jar?

A Mason jar is the vernacular term for a glass jar made especially for home canning. It's a brand name of sorts. The term comes from a type of glass jar that was patented by John L. Mason in 1858. There are many other brands of canning jars - Atlas, Ball, Sure Pak, Kerr, just to name a few. A Mason jar is to canning jars what Kleenex is to facial tissues.

Is It Safe To Can Food Without Salt?

Yes. Salt is used only to enhance flavor. In fact, one of the great bonuses of home canning is the ability to custom-design foods for people who wish to restrict their sodium intake. Commercially canned food products have a tremendous amount of salt added to them. What's more, over the last 20 years, there has been a significant rise in the amount of sugar and corn syrup added to commercially canned foods. Salt, sugar, and corn syrup are added to many manufactured foods to conceal the taste of inferior-quality products.

In The Old Days, People Didn't Use A Pressure Canner. Nobody Got Sick

Nonsense. People got sick all the time. In those days, the risk of botulism was ever-present. I'll bet your great-grandma can recall stories of entire families and church socials being sickened by botulism.

The fact of the matter is that, at one time, low-acid foods were processed without a pressure canner. Back then, food was cooked until it was almost mush before it was canned and then sealed up. In those days, canning jars were filled and placed in a boiling water bath canner for 3 hours. The resulting food was of poor quality. But mushy, uncertain food was a whole lot better than starving to death during the winter months. The risk of contaminated food is the reason you can still find canning recipes that recommend all home-canned foods be boiled for 10 to 15 minutes at a full rolling boil before serving. Canning guides caution feeding spoiled food to animals. That's because botulism can sicken or kill Fido just as fast as it can sicken or kill you. Using a pressure canner properly and according to a manufacture's specifications will give 100% bacteria kill in low-acid foods.

Must Jars Be Sterilized Before Canning?

No. The jars only need to be clean.

Why is Liquid Sometimes Lost from Glass Jars During Processing?

The loss of liquid from jars during processing occurs for a few different reasons. Liquid can be lost because the temperature in the pressure canner was fluctuating. Liquid can be lost if the jars were packed too full with product or if the food was not heated before packing.

Air bubbles in the jar or a sudden loss of pressure in the canner will sometimes cause liquid loss. Sometimes, starchy foods such as corn will absorb the liquid in the jar. It is not necessary to replace the lost liquid.

The Complete Granny Miller

Occasionally, the food will darken in the jar, but it is still safe to eat.

Why Do My Peaches and Tomatoes Float In The Jar?
Fruit will float in a jar if it was packed too loose or the syrup was too heavy. It is also possible that air remaining in the cell tissues of the fruit will cause the fruit to float.

Can I Use My Pressure Canner For Processing Fruits And Tomatoes?
Yes. You can do it one of two ways.
Method 1
If your pressure canner is deep enough, you can use it like a water bath canner. The important thing is that the water completely covers the tops of the jars by at least 1 to 2 inches. With most models of pressure canners, this is easily accomplished with half-pint and full-pint jars. If your pressure canner isn't deep enough to use like a water bath canner, you can still use it.
Method 2
You may use a pressure canner to process fruit and tomatoes at 10 pounds of pressure for altitudes of 1000 ft. sea level or lower. See the charts in the previous section for adjustments for altitudes higher than 1000 ft. sea level. For tomatoes and fruit process:
- Pints for 20 minutes
- Quarts for 25 minutes

There Are Black Spots on the Underside Of TheLid. Is The Food Safe To Eat?
Yes. Sometimes, naturally occurring compounds in food will leave a deposit. It's harmless.

How Can I Tell If A Jar Has Sealed?
If you are using a modern two-piece lid and band closure, the lid will be pulled down tightly on the jar after the jar has cooled. A modern sealed lid will have a slightly concave profile. You can test the lid by pressing your finger into the center of the lid. There should be no give to it. If the lid springs up and down, the jar has not sealed. You can also test the seal by lifting the jar by the rim. If a recently processed canning jar has not sealed, the food should be promptly eaten, frozen, or reprocessed for the full amount of time.

Can I Reuse Canning Lids?
No. It's not recommended to reuse one-trip lids. That said, some people do reuse one-trip lids with a varying amount of success. If you do reuse one-trip lids, expect losses over time and in storage due to improper sealing. There are canning lids that are made especially to be reused. One brand is Tattler, and I recommend them. You will need to follow the special instructions when using Tattler lids because they are applied differently from regular one-trip lids.

Tattler Reusable Canning Lids

Tattler lids are a modern reusable two-piece lid system for home canning. The lids consist of a white plastic disk and a red rubber ring or gasket. They are completely BPA-free. The two-piece system uses a standard modern jar band to keep the lid and gasket securely in place on the jar while the jar is being processed in the canner.

Tattler lids work in principle a little like the old-time zinc lids or wire bail jars.You may be too young to remember, but old-time canning jars used a two-piece lid system. A rubber ring was first attached to the lip or shoulder of a jar, and then a zinc lid or glass top was attached to the jar with a bail wire assembly or screw threads. In those days, instead of tightening the bail or zinc lid firmly onto the jar before processing, the lid was left loose or the bail wire was left up. This allowed the food in the jar to vent while it was inside the canner. After processing time was complete, the jars were removed from the canner, and the lid or the wire bail was tightened or clamped down immediately. This was so that a vacuum could be formed inside the jar.

Reusable canning lids are the ultimate in sustainability and semi-self-reliance. Tattler lids are much more expensive than regular one-trip lids. But they will easily pay for themselves over time. I don't use Tattler lids for all of my canning needs. But I do keep at least half of my canning jars coupled with reusable canning lids. That's because I'm old enough to remember the canning lid shortage of 1975 - 1976, and I don't ever want to go through that again if I can help it.

Tattler lids can be more unforgiving than one-trip lids. Greater care and attention need to be employed when using them.

To use Tattler Lids, first scald the lids and rubber rings in hot water. Keep the lids and rings hot until ready to use. Fill the canning jar with food and then wipe the rim of the jar to ensure a good, clean surface for sealing. Place the lid and rubber ring combination on the jar and apply a band. Tighten the band. Next - and this is vital - **loosen the band just a little**. Process the food according to the chosen recipe. When the processing time is completed, remove the jars from the canner. Then, while using hand protection, immediately and firmly tighten the band down on the jar. Allow the jars to cool undisturbed for 8 to 12 hours. After the jars have completely cooled, remove the metal band. Check the seal by gently lifting the jar by the rim of the lid. The contents and date of the food may be marked with a wax pencil or a piece of freezer tape and a pen.

To remove the lid, gently insert a butter or table knife between the rubber ring and the jar. Never use a sharp knife. Wash the plastic lids and rubber rings and store them in a cool, dry location until needed again.

Broken Canning Jar

I seldom get a canning jar break while in a pressure canner or a water bath canner. It does happen, but not too often. Sometimes, during processing, you can hear a jar break in the canner. But sometimes you can't. And often when you're least expecting it, a nasty surprise and a big mess are in store for you when the top of the canner comes off.

There are several different reasons for jar breakage that you might want to keep in mind for your canning.

Using one-trip glass jars for canning is asking for trouble. A one-trip jar is an older type of glass container that was used for peanut butter, mayonnaise, or other commercial food products. At present, it's hard to buy food in glass jars, so it's not too much of a problem anymore. But sometimes, when buying a box of old canning jars at a yard sale or auction, you will come across them. Be sure to check all second-hand jars to ensure they are indeed canning jars. Sometimes a true Ball canning jar will appear to be a one-trip jar, but it really isn't. Check to make sure by turning the jar upside down and look for a 'B' on the bottom.

Canning jars can break in a pressure canner if you use a metal utensil, like a fork or a knife, to remove air bubbles. Metal utensils can cause invisible interior scratches in a jar. These scratches or small fissures weaken the jar and make it susceptible to breakage during pressure canning. Using steel wool or a wire brush to clean jars, or using metal utensils to remove the food contents from a jar, or scraping the inside of a jar with a knife can all cause interior scratches.

- Jars can break during processing by putting hot food into a room-temperature or cold jar
- Jars can break by putting a room-temperature or cold jar into boiling water.
- Jars can break by putting frozen or incompletely thawed food into a canning jar.
- Jars can break when removed from the canner if they are placed directly on a cool countertop or wet surface.
- Sometimes canning jars will break during water bath canning if they are not secure in a rack. With water bath canning, unless the jars are well seated in a rack, the jars can knock against each other and break if the water boils too rapidly.

Why a Canning Jar Lid Will Come Unsealed

A canning jar lid failure is something I'll come across once in a blue moon. It's a relatively rare occurrence. When the vacuum inside a canning jar is breached, bacteria and yeast begin to grow inside the jar, causing fermentation and gas. Sometimes the force of the gas from the fermentation will work to crack a lid off a canning jar. The resulting lid failure is usually obvious with slime and mold on top of the food product. But sometimes there is no visible sign of food spoilage. But that doesn't mean the food is safe to eat. There are many reasons for faulty seals on home-canned foods.

- Sometimes jars will not seal properly if the rim of the jar has not been wiped perfectly clean before the lid and band are applied. If grease or other materials from the food product are forced under

89

the lid during processing, the lid may not form a strong seal.

- Not testing seals. It is important to always remove the bands from newly processed and cooled jars and to do a lift test to ensure a solid seal. A lift test is done after the jars have been removed from the canner and cooled for 8 to 12 hours. When the jars are cooled and the bands removed, lift the jar about 2 inches above a well-padded table or countertop by the rim of the lid. If the lid holds, the seal is good.
- Reusing one-trip lids can cause lid failure.
- Canning jars that have not been kept hot before packing or suffer inadequate processing time will sometimes cause a lid failure. This is especially true when processing cold-pack food.
- Sometimes there is an unseen hairline crack in the jar or a nick on the rim.

Often, the reason for a jar with a bad seal will remain unknown. It can be a guessing game. But I do take note if a particular Mason jar repeatedly fails to seal. That jar gets retired from the kitchen and goes on to a new career in the garage or elsewhere.

Crystals in Canned Grape Juice

On occasion, when opening a jar of home-canned grape juice or grape jelly, you will find small, sharp crystals inside the jar or in the actual food product. These crystals are tartrate crystals and are formed by the naturally occurring tartaric acid in the grape juice. The crystals are perfectly harmless and in no way affect the safety of the food product.

Tartrate crystals in grape products are formed by sediment in grape juice or other grape-based products like wine or jelly. Many canners and jelly makers don't like them, and often, novice canners are upset to see them sitting at the bottom of a jar of grape juice.

To prevent tartrate crystals in home canning, allow the grape juice to rest overnight in a refrigerator, allowing all the sediment to collect at the bottom of the container. In the morning, carefully pour off only the clear juice and try not to disturb the sediment. By keeping the grape sediment out of the juice before it is canned, clear grape juice without little crunchy things floating around is almost guaranteed. And a beautiful, sparkling blue ribbon jelly is within reach.

In nature, grapes are the richest source of tartaric acid. The kitchen aid known as Cream of Tartar is obtained from grape sediment and is made from by-products that are left over from winemaking. All grapes contain tartaric acid.

But some have more than others. Foods and wines from Concord grapes are notorious for forming crystals. So, if you should ever happen upon crystals in your home-canned grape jelly, or grape juice - don't be alarmed. They're natural.

The Complete Granny Miller

How to Can & Freeze Green Beans

Green beans and yellow wax beans are a favorite summertime vegetable for many people. They are easy to grow and easy to can or freeze.

Keep in mind when growing green beans that they need plenty of good air circulation. Green beans shouldn't be weeded or tended to while they are wet or damp. Working around green beans while they are wet will spread fungus and disease.

Selecting Green Beans for Canning

Always try to obtain the freshest and most tender green beans possible. Reject beans that are over-mature, hollow, tough, limp, or floppy. It takes approximately 1 ½ pounds of raw, unsnapped green beans to produce a finished canned quart.

Fresh green beans, whether picked from the garden or purchased at a farm stand or market, need to be washed and trimmed before canning or freezing. Trimming or snapping beans is easy, but it is time-consuming.

"Many hands make light work."

Snapping or cutting beans goes much faster with good company and conversation.

The stem end of the bean needs to be removed. The bean is snapped or cut in half or into thirds. Snapped beans are faster. But cut beans look better. Either is fine. It just depends on your preference. After the beans are cut or snapped, place them into a sink of cold water to rinse and clean them. You may have to swish or rub the beans gently with your hands to remove any clinging dirt or grass. Drain the sink and rinse the beans well under running cold water. Remove the beans to a bowl. The beans are now ready to be canned or frozen.

Get Ready

Green beans are a low-acid food. The only dependably safe method for home canning low-acid foods is with a pressure canner. Green beans can be pressure canned by using either the raw pack or hot pack method. I prefer the raw pack method because it saves time and energy. The raw pack method is presented here.

The hot pack method is essentially the same as the raw pack method, with the key difference being that the beans are cooked or blanched before being packed hot into a hot canning jar. The advantage of a hot pack is that it allows more beans to be packed into a jar, and the beans don't tend to float. The disadvantage of the hot pack method is that the beans can become mushy if cooked for too long.

Usually, green beans are canned with salt for added flavor. Salt is not a necessary ingredient for successful home canning. People on low-sodium diets may prefer to skip the salt altogether. If you choose to add salt, the standard measure is:

- 1 teaspoon for Quarts
- ½ teaspoon for Pints

I use half those amounts in my beans.

The Complete Granny Miller

Only prepare enough green beans for one canner load at a time.

Gather and assemble the jars, lids, bands, jar lifter, funnel, and the pressure canner, well before canning day. Check to make sure that all equipment is in good working order. Nothing is worse than stopping in the middle of canning to go and hunt for something you forgot. Or worse, having to make a run to the hardware store to replace equipment or parts that don't work correctly.

Visually examine all jars and rims for cracks, nicks, or sharp edges. Examine the pressure canner and gasket carefully. Wash the jars and bands in hot, soapy water and rinse well. Dry the bands and set them aside. Keep the jars hot. Place the rack in the pressure canner and add the recommended amount of water according to your canner's manufacturer. If you live in a hard water area, add 1 tablespoon of white vinegar to prevent hard water marks on the canning jars.

Canning Green Beans

Begin to heat the canner and a kettle of water while you are working. The water in the kettle will be poured into the jars to cover the beans. Simmer lids for 3 to 5 minutes if necessary, and then keep hot until ready to use.

* Since the first publication, canning lids no longer need to be simmered – just washed or rinsed. Follow the manufacturer's directions on the box*

Fill a hot jar with green beans. Pack the beans tightly into the jar, but don't cram or overstuff.

Pour boiling water into the jar and over the beans. Leave a 1-inch headspace in the jar. A jar funnel is an easy way to help you determine headspace. The distance from the bottom of the jar funnel as it sits inside a canning jar is 1 inch.

It is necessary to remove air bubbles and air pockets from the jar. Do this by sliding a non-metallic object or spatula down the sides of the jar. You will see tiny air bubbles rise to the top of the jar, and the beans will float a little. This is normal.

Wipe the rim of the jar and the jar threads with a clean, damp cloth. The rim must be perfectly clean and free of food particles so that the lid makes firm contact and a good, strong seal. Place and center a canning lid on the jar with the sealing compound next to the jar rim. Screw the lid band down evenly and firmly. But don't over-tighten.

Place the sealed jar into the canner and fill the remaining jars one at a time, placing them into the canner until the load is complete. The water in the canner should be hot and simmering. When all the jars are filled and the load is complete, put the lid on the pressure canner and close it.

Heat the canner with the pressure control weight off. Heat it until a steady stream of steam comes out of the vent. Allow the steam to vent from the canner for about 7 to 10 minutes or according to your canner manufacturer's directions.

It is essential to drive all of the air out of the pressure canner, especially if you are using a dial gauge. Air pressure and steam pressure together may give a faulty reading on a dial gauge. Once the canner has been properly vented, apply the control weight.

Processing time for green beans at 10 pounds of pressure at 1000 ft. of sea level or less is:

- 25 minutes for Quarts
- 20 minutes for Pints

You **must** make a pressure adjustment for higher altitudes and elevation.

Adjustment for Pressure Canner

Altitude and Elevation Adjustment Needed for Pressure a Pressure Canner		
Altitude In Feet	Dial Gauge	Weighted Gauge
0 – 1000 ft.	10 psi	10
1001 – 2000 ft.	11 psi	15
2001 – 4000 ft.	12 psi	15
4001 – 6000 ft.	13 psi	15
6001 – 8000 ft.	14 psi	15
8001 - 10,000 ft.	15 psi	15

- With a dial gauge, the processing time is counted from the time the proper pressure is reached.
- With a control weight, the proper pressure is counted from the time the weight first begins to jiggle or rock. A jiggle of 1 to 4 times a minute is about right for a control weight canner.

With a dial gauge, the aim is to keep the heat steady and even so that the pressure remains stable. Adjust the heat on the stove if necessary to keep an even and correct pressure throughout the entire processing time. It may take some trial and error to determine the heat setting on your stove that keeps the pressure steady. If at any time the pressure goes below the recommended amount, bring the canner back up to the correct pressure and begin the timing from the beginning.

Once the processing time is complete, turn off the heat under the canner and let it stay on the burner to cool. Alternatively, you can carefully remove the canner from the heat and allow it to cool naturally. Do not try to hasten the cooling process by using cold water or a fan.

Only after the pressure in the canner has returned to normal pressure is it safe to open the lid. For a control weight canner, the pressure will have returned to normal when the vent stops hissing. If the control weight hisses at all when you touch it, it indicates that too much pressure remains inside the canner. Leave it alone and give it more time to cool.

For a dial gauge canner, when the gauge reads '0', it is safe to open the canner. Take care and use caution while opening the canner lid. All surfaces of the canner will be extremely hot. Always open the canner with the lid facing away from you and allow the hot water to run off and drip back inside the pot.

Carefully remove the hot jars from the canner with a jar lifter. Place the jars

The Complete Granny Miller

on a thick towel, thick newspapers, or on a wooden board well out of the way of drafts. Leave the jars undisturbed for 8 to 12 hours.

It is normal for the food or liquid inside a canning jar to be boiling when it is removed from a pressure canner. Frequently, a pinging or snapping sound is heard as the jars begin to cool. The pinging sound is the lid being pulled down. It indicates that a vacuum seal has been achieved.

After the jars have thoroughly cooled, it's critical to check the seal. Remove the screw band before checking the seal. Check the seals by gently lifting the jars by the lid 2 inches over a padded surface. For one trip, canning lids notice if the center of the lid has been pulled down to create a slightly concave surface. If you are not sure if a jar's lid has been pulled down, push into the center of the lid. If the center of the lid moves up and down, the jar has not sealed. The food should be reprocessed within 24 hours, or frozen or eaten promptly. Jars should be wiped clean if necessary, labeled, dated, and stored in a cool, darkplace.

Freezing Green Beans

Green or yellow beans can easily be frozen. Freezing results in beans that have better color retention and a fresher taste. Frozen green beans are much preferred over canned beans in many recipes. Preservation by freezing does not sterilize food. It only retards the growth of microorganisms and slows down enzyme activity and oxidation. Green beans, like most vegetables, must be scalded or blanched before freezing. Scalding cleanses the surface of dirt, reduces the action of enzymes, and brightens the color of vegetables. Without proper blanching or scalding, vegetables will begin to deteriorate in the freezer after about four weeks.

How to Freeze

Prepare the green or yellow beans the same as for canning. Bring a large pot of water to a rapid boil. Place the snapped or cut beans into the boiling water for 2 to 3 minutes. After 2 or 3 minutes, remove the beans from the boiling water and rapidly cool them by placing them in a sink of cold water and ice. Once the beans are cool, drain them well. Package the beans into rigid freezer containers, freezer bags, or Vacu-Seal bags. Freeze immediately. Frozen green beans will last between 12 to 18 months in a freezer under good conditions.

How to Can and Freeze Sweet Corn

Sweet corn, like most vegetables, is a low-acid food. Low-acid foods are only safely and dependably home-canned with a pressure canner. I use the raw pack method when canning corn because it saves time and energy. It's the method I'm going to share with you here.

Before canning day, gather and assemble the jars, lids, bands, jar lifter, jar funnel, a non-metallic tool to release air bubbles, and the pressure canner. Check to make sure that everything is in good working order. Visually examine all the jars. Check the rims of the jars to make sure that there are no cracks, nicks, or sharp edges. Wash the jars, lids, and bands in hot soapy water. Dry the bands and set aside. Keep the jars hot until ready to fill with corn.

Begin heating a large kettle of water. The water in the kettle will be poured into the jars and over the corn. Put the recommended amount of water into the pressure canner and begin to heat it. Try to time the hot water in the kettle and the heating of the canner to all coincide with when you will be ready to pack the jars with corn.

Prepare the Sweet Corn

Collect fresh-picked sweet corn. Don't try to do more than one canner load of sweet corn at a time. Corn is especially perishable and time sensitive. Husk the corn and remove the corn silk. To make kitchen clean-up easier, husking is best done outdoors. After the corn is husked, rinse the ears well with cold water. While working quickly and without delay, cut the corn kernels from the cob and place them into a bowl. Sweet corn will turn sour if it waits too long before being packed into jars after being cut off the cob. Do not scrape the cobs.

By now, the water in the canner should be hot, and the kettle of water should be simmering. If you choose to add salt to your corn, the standard measure is:

- 1 teaspoon of salt for Quarts
- ½ teaspoon of salt for Pints

I use half those amounts in my corn.

Fill a hot jar with corn, taking care not to pack it in too tight. Next, pour boiling water into the jar and over the corn, leaving a 1-inch headspace. Slide a non-metallic object down the sides of the jar to release any trapped air bubbles. Wipe the rim and sides of the jar with a clean, wet cloth, then apply a lid and a band.

Place the jar into the canner so it will remain hot while you work filling the other jars. Keep the water in the canner at a simmer while you are working. When all the jars are filled and you have a full canner load, put the lid on the canner and close it.

Turn up the heat and allow the pressure canner to vent for 7 to 10 minutes or according to your canner's manufacturer's instructions. After proper venting, place the weighted gauge on the canner. Processing time for a weight gauge pressure canner begins to be counted from the moment the weight first begins to jiggle or rock. Processing time on a dial gauge canner begins when full pressure has been reached.

Corn is processed at 10lbs. of pressure for:

- 55 minutes for Pints
- 1 hour and 25 minutes for Quarts

Don't forget to make an altitude adjustment in the processing time or pressure if you are above 1000 feet sea level.

The Complete Granny Miller

Altitude and Elevation Adjustment for Pressure a Pressure Canner		
Altitude	Dial Gauge	Weighted Gauge
0 – 1000 ft.	10 psi	10
1001 – 2000 ft.	11 psi	15
2001 – 4000 ft.	12 psi	15
4001 – 6000 ft.	13 psi	15
6001 – 8000 ft.	14 psi	15
8001 - 10,000 ft.	15 psi	15

When the processing time is complete carefully remove the canner from the stove or turn off the heat under the canner. In either case, allow the canner to cool naturally. The canner is safe to open when the dial gauge reads '0' or the weighted gauge ceases all hissing. Always open the canner lid facing away from you. Remove the jars from the canner with a jar lifter. Place the jars on a towel, a board, or on a thick layer of newspapers, well out of the way of drafts. Allow the jars to cool undisturbed for 8 to 12 hours.

When the jars have completely cooled, check the seal. Remove the screw band before checking the seal. The seal is checked by gently lifting the jar by the lid or pushing down into the center of the lid. The lid should be slightly concave and have no spring to it. If the lid bounces up and down, the jar has not sealed, and the corn should be frozen, reprocessed, or eaten promptly. Jars should be wiped clean, labeled, and stored in a cool, dark place.

Hints for Working with Sweet Corn

Sweet corn is messy and sticky. It squirts everywhere when you cut it off the cob. Be sure to have plenty of wet cloths while you work to wipe your hands. You'll probably have to clean off your knife a couple of times. Do not work near a window or wall if you can help it. If you do, you'll be cleaning windows and washing walls when you're done. Don't say I didn't warn you.

With corn and other starchy foods, sometimes liquid is lost in the canning jar. Don't be concerned. As long as the seal is good, the food will not spoil. The food may darken a little on the top over time, but it is still safe to eat.

Freezing Sweet Corn

Freezing sweet corn is much simpler than canning sweet corn. It results in a superior product that is fresher tasting. When leftover cut corn isn't enough to make a full canner load, I always freeze it. Corn for freezing is prepared the same way as for canning: cut quickly from the cob and handled in small amounts.

There are several different ways to freeze corn. Here's what works for me: Place the cut corn in a saucepan and add a small amount of water to it. Add just enough water to barely cover the corn. Heat the corn over medium heat for about 4 or 5 minutes. You want the corn just to begin to simmer and partially cook. The corn will probably thicken a little, and that's okay. After 4 or 5 minutes, drain the corn into a colander, and then add ice to the corn in

The Complete Granny Miller

the colander. When the corn has cooled, you may need to remove any remaining ice. Pack the cooled corn into rigid containers, freezer bags, or Vacu-Seal bags. Frozen corn will keep for 12 to 15 months, depending upon freezer conditions.

A Guide to Canning and Freezing Tomatoes

Most heritage tomatoes contain enough naturally occurring acids that they can be safely home-canned and processed in a water bath canner without the addition of lemon juice. But many modern varieties of tomatoes don't contain enough naturally occurring acids. Those types of tomatoes should be processed with added lemon juice to ensure acidity. When in doubt, always add lemon juice or vinegar to tomatoes before canning them. When canning tomatoes, select only firm, red ripe tomatoes and avoid tomatoes that are too ripe. That's because the acidity of tomatoes declines the further they ripen. Also, keep in mind that extremely soft or excessively ripe tomatoes can turn into a mushy, overcooked mess during processing if you're not careful. In my opinion, overripe tomatoes are better suited for making juice or tomato sauce.

Get Ready

Before canning day, collect and assemble the jars, bands, lids, jar lifter, a non-metallic bubble release tool, large bowls, knife, a wooden spoon, and the waterbath canner. Visually examine all jars and rims for cracks, nicks, or sharp edges. Wash the jars and bands in hot, soapy water. Dry the bands and set them aside. Keep the jars hot until ready to fill. A dishwasher is good for keeping jars hot. But a sink or dishpan full of hot water works just as well. As does pouring boiling water into the jars and standing them in a sink or shallow pan until ready to use.

Wash and Sort the Tomatoes

Prepare only one canner load of tomatoes at a time. Preparing the tomatoes will take a bit of time, depending upon the size of the jars and the size of the canner load you intend. Fresh collected tomatoes should be washed in cool, soapy water and then rinsed well to remove any garden dirt. This step helps to sort the tomatoes out by size and can be done well ahead of canning day.

Peeling the Tomatoes

Tomatoes will need to be peeled before they are canned. If you don't peel the tomatoes, the skins become hard bits that float in the sauce or other tomato products.

Peeling tomatoes is a messy job, but thankfully, an easy one. It will take sometime to peel the tomatoes. Try to coincide heating the water in the canner with simmering the jar lids (if you are using lids that need simmering), for when you're finished peeling the tomatoes. It may take some multi-tasking to get the timing just right. The important thing is to have the water in the canner simmering hot by the time you fill the jars. Here's how to peel and remove the skin from tomatoes.

To remove the skins from tomatoes, place a few tomatoes of about the same size into a pot of rapidly boiling water.

The Complete Granny Miller

Keep the tomatoes in the pot until their skins begin to crack. It usually takes about 3 minutes for the skins to split. Sometimes it will take longer. But take care not to keep the tomatoes in the boiling water for too long, or you'll end up with hot tomato mush.

After the skin has split, remove the tomatoes from the boiling water and place them directly into a sink or large bowl of ice-cold water and ice cubes. This will rapidly cool them down. When cool, peel the skin away from the tomatoes and remove the core and any green parts.

Small tomatoes may be kept whole. But large tomatoes should be quartered or cut into smaller pieces so that they will fit into a jar. After the tomatoes have been prepared and the water in the canner is hot, it's time to fill the jars.

Put Tomatoes in a Jar

Salt is optional when canning tomatoes.

The standard canning recipe measure for added salt is:

- 1 teaspoon for Quarts
- ½ teaspoon for Pints

I routinely use those amounts of salt with my tomatoes. For extra safety or if you are unsure about the acidity of the tomatoes, add 2 tablespoons of lemon juice per quart of tomatoes, and 1 tablespoon of lemon juice per pint. I usually use bottled lemon juice.

Next, fill a hot jar with tomatoes, leaving a ½ inch headspace. It may be necessary to squeeze and push the tomatoes down while filling the jars to release more juice from them. The juice helps to achieve the proper headspace. I have a small wooden stomper that I use for this purpose. But a wooden spoon works just fine.

Tomatoes tend to collect air pockets and bubbles in the jar, and it's essential to release them. I do this by running a non-metallic object or spatula down the sides of the jar. After the air bubbles have been released, wipe the rim and threads of the jar with a clean, damp cloth.

When working with tomatoes, extra care needs to be taken when cleaning the rim and threads of the jar. It is easy to miss a seed or a small piece of tomato pulp. If a seed gets between the rim of the jar and the lid, the jar will fail to seal. After the jar rim and jar threads are perfectly clean, apply the lid and the screw band. Don't over-tighten the band.

Place the filled jars into the simmering canner and turn up the heat so the water can begin to boil. With water bath canning, it is imperative that boiling water completely covers the jars by 1 to 2 inches. If more water needs to be added to the canner, ensure it is boiling water. To add extra water, pour boiling water along the sides of the canner and not directly over the jars.

Processing time for water bath canning is counted from the time the water begins to boil <u>AGAIN</u> after the jars have been loaded.

A gentle but steady boil is what you're trying for.

Processing time for tomatoes for altitudes at 1000 ft. sea level or less is:

- 45 minutes for Quarts
- 35 minutes for Pints

The Complete Granny Miller

If your location is above 1000 ft. above sea level, you will need to adjust the processing time.

Altitude and Elevation Adjustments Needed for a Water Bath Canner	
0 - 1000 ft.	No adjustment
1001 – 3000 ft.	Add 5 minutes
3001 – 6000 ft.	Add 10 minutes
6000 – 8000 ft.	Add 15 minutes
8001 ft. - 10,000 ft.	Add 20 minutes

Once the processing time is completed, remove the jars immediately from the canner and place them on a heavy towel, wooden board, or thick layer of newspapers. Keep the jars well out of the way of drafts. Leave the jars undisturbed for 8 to 12 hours.

After the jars have cooled, it is essential to check the seal. Remove the screw band before checking the seal. The jar seal is checked by gently lifting the jar over a padded table or counter top by its lid, or pushing down into the center of the lid. A sealed lid should be slightly concaved and have no spring to it. If the lid bounces up and down, the jar has not sealed. The tomatoes should be frozen, reprocessed, or promptly eaten. Often, the sides of the jars will be sticky and need to be wiped clean. As with all home-canned foods, label with the contents and the date, and store in a cool, dark place.

Hints For Canning Tomatoes
Pressure canning tomatoes can save time and fuel. When tomatoes are pressure canned, the lemon juice can be reduced by half.

Every summer, I manage to get a couple of jars of tomatoes that don't seal. Most often, it's caused by seeds or bits of tomato being pushed up under the lid during processing. When I have a jar of tomatoes that doesn't seal, I freeze the tomatoes by emptying the contents of the jar into a rigid freezer container or a freezer bag.

Often, tomatoes and other fruits or vegetables packed by the raw pack method will tend to float in a canning jar. This is normal and is no cause for concern. Raw vegetables, fruits, meat, and other foods are denser when they are raw than when cooked. When raw food is packed into a canning jar, it takes up more space than hot or partially cooked food. The floating occurs because the vegetable cooks during the canning process, resulting in shrinkage. Floating fruit or tomatoes can be overcome by heating or semi-cooking the food product before it is packed into the jar. Heating food before filling a canning jar is known as 'hot pack,' and many people prefer it when canning. The advantage of a hot pack is that you can fit more food into a jar, which means you don't need as many jars. This is good to know if you are short on jars. However, if you heat or cook tomatoes before filling the jars, they do tend to fall apart and are harder to handle. Raw packing tomatoes results in a superior product, which is why most people use it.

The Complete Granny Miller

Freezing Tomatoes

You can freeze tomatoes if you don't care to be bothered with canning them or don't have the equipment. Prepare the tomatoes exactly as if you were going to can them: wash, peel, and core them. But instead of canning them in Mason jars, they'll be placed directly into freezer bags or rigid containers. Frozen tomatoes are wonderful in chili or other dishes and have a much fresher taste. They make for an excellent Cream of Tomato soup. When frozen tomatoes are defrosted, there will be a separation of the tomato flesh and liquid. To manage this, you can pour off the liquid and have a dryer tomato product. Or you can turn the freezer bag or container upside down a couple of times, and the tomatoes will reincorporate.

At the beginning and the end of the tomato season, I usually freeze tomatoes. That's because when tomatoes first come on in the garden, typically, there isn't enough to make a full canner load. Freezing rather than canning tomatoes is a much more sensible and energy-saving option if you have just a few pounds of tomatoes to work with.

How to Can Mushrooms

Home canning fresh mushrooms is easy. Like all low-acid vegetables (mushrooms are actually fungi), mushrooms must be canned with a pressure canner. Mushrooms are usually packed in half-pint or full-pint jars. It takes approximately 1 half a pound or 8 oz. of fresh mushrooms to fill a half-pint jar. When selecting mushrooms for home canning, it's best to choose only firm, bright, and small to medium-sized domestic mushrooms. You can tell if a mushroom is fresh by its clean color and an unopened veil or cap. Reject mushrooms that are old, darkened, bruised, or imperfect in any way.

Gather What You'll Need

Gather and assemble all canning equipment ahead of time. Make sure your pressure canner is in good working order. Wash the jars in hot, soapy water. Rinse them well and keep them hot. Begin to heat the pressure canner and a kettle of fresh, clean water. The water in the kettle will be poured into the jars and over the mushrooms.

Trim mushroom stems and any discolored areas. Slice medium-sized mushrooms evenly, but leave the smaller mushrooms whole. Soak mushrooms in cool water for about 3 minutes to remove any dirt that may cling to them. Rinse well in cool water.

Next, place the mushrooms in a large pot and cover with water. Bring the pot to a gentle boil for about 3 to 5 minutes. Remove mushrooms from the heat; drain the mushrooms, and begin filling hot jars.

Salt may be added if desired. I use salt when canning mushrooms and recommend:

The Complete Granny Miller

- ⅛ of a teaspoon for Half Pints
- ¼ of a teaspoon for Full Pints

But be aware that some people prefer twice as much salt for canned mushrooms.

Pour hot water from the kettle over the mushrooms, leaving a 1-inch head space. Remove any air pockets by sliding a non-metallic object along the sides of the jars. Wipe the rims of the jars clean and apply lids and bands. Place the jars in the pressure canner. Close the lid and increase the heat under the pressure canner until steam begins to vent. Allow the canner to vent for 7 to 10 minutes, or as per your canner's manufacturer instructions. Once the pressure canner is vented correctly, apply the weighted gauge to the canner. Bring the canner to a full pressure of 10 pounds.

Processing time for both half pints and full pints is 45 minutes
Remember, processing time is counted from the time full pressure is achieved. If you live at or above 1000 ft. of sea level and are using a weighted-gauge canner, increase the weight to 15 pounds and keep the processing time the same.

If you are using a dial-gauge pressure canner and live at an altitude over 1000 ft. of sea level, keep the processing time the same but increase the pressure as follows:

Altitude and Elevation Adjustment for Pressure Canner		
Altitude	Dial Gauge	Weighted Gauge
0 – 1000 ft.	10 psi	10
1001 – 2000 ft.	11 psi	15
2001 – 4000 ft.	12 psi	15
4001 – 6000 ft.	13 psi	15
6001 – 8000 ft.	14 psi	15
8001 - 10,000 ft.	15 psi	15

After processing time is completed, remove the pressure canner from the heat and allow the pressure to return to '0' naturally. Do not hurry the cooling down. Open the canner carefully, keeping the lid away from your face. Remove jars with a jar lifter and place them on a dry, padded surface or wooden board. Leave the jars undisturbed for 8 to 12 hours. Remove bands and check seals.

Mark the jars with the date and store them in a cool, dry location.

Home Canning Meat
Rabbit•Chicken•Beef•Pork•Lamb &Small and Large Game

Chickens and rabbits are usually the first meat animals that a new homesteaderor garden farmer will acquire. They are small animals, quite manageable, and are a good fit with backyard gardens. I'm actually surprised that more people don't keep them. Rabbits and chickens require less daily care than one or two neurotic house cats.

While not all towns or municipalities permit keeping chickens, many people find ways to raise rabbits for food. The fact is, your neighbors don't need to know that the rabbits in your basement or garage are for food and are not pets.

Rabbits reproduce quickly and can be harvested with little trouble several times a year. By canning rabbit, you can meet some of your food storage goals. For the most part, rabbit and chicken recipes are interchangeable. Except that a rabbit has less fat than a chicken, so keep that in mind for recipes when you are cooking a fresh rabbit. Like chicken, if you cook rabbit too fast, the meat can end up tough and stringy. Except for frying, use the slower methods when cooking rabbit. When using canned rabbit or chicken, you don't have to worry about tough meat. That's because the meat was already pressure-cooked and is tender and moist.

Canned chicken and rabbit are tremendous time savers. I use canned chicken and rabbit in salads, casseroles, barbecue, in white gravies and sauces, over biscuits, and in any recipe that calls for cooked chicken. Chicken pot pie is one of my favorites, as is rabbit stew. You can interchange rabbit and chicken in canning recipes. So, a recipe for chicken soup becomes rabbit soup. White chili made with rabbit or chicken is good and is easily canned. Pick any recipe that calls for chicken and use rabbit instead. The variations in recipes are endless and depend only upon the cook's imagination and ingenuity.

There are a couple of different ways to home-can beef, pork, lamb, chicken, rabbit, and game meat. You have a choice between the 'hot pack' and the 'raw pack' methods. There is also a choice between the 'bone-in' and the 'bone-out' method. I think the best way to can freshly harvested rabbit or chicken is with the hot pack bone-out method.

Hot pack bone-out produces a product that is ready to use right off the pantry shelf with liquid for gravy or sauce. It is the method that I most often use when canning whole rabbits or whole chickens. When the bones are left in during canning, the flavor of the meat is stronger. I don't notice it too much with rabbits, but it is quite noticeable with chickens or squirrels. The difference in flavor is not a bad difference. It's just different. To me, it's like the difference between mild, white meat chicken and really dark meat chicken. In certain recipes, I don't care for the stronger flavor from the bone-in method. The bone-in method is most often used in canning when it may be too much trouble to remove the bones.

Canned meat on a pantry shelf is a quick and convenient food. When I'm in a hurry or pressed for time, I just want to open the jar and cook. I'd rather do the work boning chicken or rabbit while I'm canning and not later when I'm in a hurry and cooking. No matter which method is chosen, canning meat with the bone-in or bone-out is a pretty simple affair.

The Complete Granny Miller

Equipment

You'll need a working pressure canner. You'll also need canning jars, lids, and bands; a jar lifter, hot mitts, and ordinary kitchen equipment - bowls, knives, towels, etc. I find that wide-mouth Mason jars work best when canning meat. That's because wide-mouth jars are easier to fill, empty, and clean. A wide-mouth jar is easier to pack, and this is especially important when canning meat with the bone left in. When packing bone-in chicken or rabbit into a canning jar, you may need to adjust the pieces to fit without wasting space. There is no sense packing just one chicken leg into a quart jar when you could actually have fit a chicken leg, another thigh, and two wings into the jar.

Also, wide-mouth jars are easier to clean after they've been used. Sometimes the inside of a jar becomes coated with bits of cooked meat. That makes the jar hard to scrub out, even with a bottle brush. With a wide-mouth jar, it's much easier to put your hand inside the jar and scrub it clean. Boneless meat sometimes will pack into a solid, dense mass when canned. With a wide-mouth jar, removing the meat is much easier. With a regular mouth jar, it can be a real struggle to get the meat out of the jar.

Yield

The number of jars that any given amount of meat will yield varies with the manner and the method by which the jars were packed. Yields will vary due to the size of the meat pieces, whether or not a raw or hot pack was used, and whether or not the bone was left in. All are factors that determine jar yield and outcome. As a general rule of thumb, allow 2 to 2 ½ pounds of boneless meat per quart jar. When canning bone-in chicken or rabbit, plan to allow for between 2 ½ to 4 ½ pounds of meat per quart jar. The bone is heavier than you'd think.

A Word about Giblets

If you are processing a large batch of rabbits or chickens and want to can the hearts, livers, or gizzards, set them aside to be canned in separate jars. It's also a good idea to can the livers in a separate jar because the liver taste will transfer to the other giblets. I always save the livers, kidneys, hearts, and other bits when processing harvested animals. Even if I don't eat those parts, my dogs and cats will. To my way of thinking, it's wrong to waste any useful part of an animal if another animal can use it.

Hot Pack Method for Canning Bone-In

Cut the rabbit, chicken, or squirrel into pieces that will fit inside the jar. Trim off any fat. You probably won't have any fat on a rabbit, but you will have fat on a chicken, duck, raccoon, or turkey. Place the raw pieces of meat into a large pan or Dutch oven and cover with water or any type of hot broth of your choosing. The broth can be seasoned. But I would caution you to go easy on the spices and seasoning. Canning will intensify some flavors, but not for the better.

Place a lid on the pan and cook the rabbit or chicken over low to medium

The Complete Granny Miller

heat until the meat just loses its pink color when cut and tested. Pack the chicken or rabbit loosely into a hot jar, leaving a 1-inch head space. Place the big pieces in the center of the jar and fit the smaller pieces around them. Add salt if you like:

- ½ teaspoon of salt for Pints
- 1 teaspoon of salt for Quarts

Cover the rabbit or chicken with boiling hot broth and maintain a 1-inch headspace in the jar. Wipe the rim of the jar. Wiping the rim of the jar carefully is especially important with fatty poultry and some fall-harvested small game animals. Grease on the rim of the jar may prevent a seal. With rabbit or squirrel, it isn't usually a problem unless you added some type of fat to the broth.

Hot Pack Bone-Out Method
Partially cook the animal as above. Remove the pieces from the broth when they are cool enough to handle. Pick the meat from the bones and discard the skin from the chicken (unless you want to can it separately for pets). Pack the hot or warm meat into hot jars. Add salt if you like, and cover the meat with broth, leaving a 1-inch head space. Carefully wipe the rim and apply a lid and band.

Processing Time
You will notice that the processing time for bone-in meat is less than that of bone-out. This is because it takes less time for the inner core of the jar to reach 240°F when the bones are present. Bone-out meat packs solid in a jar, whereas bone-in meat does not. Whether you use the raw or the hot pack method for bone-in meat, the processing time is the same. Process jars at 10 pounds of pressure in a pressure canner for altitudes at 1000 ft. sea level or less.
You will need to adjust pressure accordingly for higher altitudes, depending on the type of pressure canner system you are using.

Altitude and Elevation Adjustment Needed for Pressure Canner		
Altitude	Dial Gauge	Weighted Gauge
0 – 1000 ft.	10 psi	10
1001 – 2000 ft.	11 psi	15
2001 – 4000 ft.	12 psi	15
4001 – 6000 ft.	13 psi	15
6001 – 8000 ft.	14 psi	15
8001 - 10,000 ft.	15 psi	15

Processing Time for Bone-In:
- Pints - 65 minutes
- Quarts -75 minutes

The processing for bone-out meat is the same for the hot pack method or the raw pack method. Process jars at 10 pounds of pressure in a pressure canner

The Complete Granny Miller

for altitudes at 1000 ft. sea level or less.

Processing Time for Bone-Out:
- Pints - 75 minutes
- Quarts - 90 minutes

You will need to adjust pressure accordingly if you live above 1000 ft. of sea level. When processing time is complete, remove the canner from the heat and allow the pressure to return to normal on its own. Don't rush the cooling process, as this may prevent jars from sealing or cause a loss of liquid. When pressure has returned to normal inside the canner, remove the jars. Place the jars on a dry towel or wooden board well out of the way of drafts. Allow the jars to cool undisturbed for 8 to 12 hrs. After the jars have cooled, remove the bands and check the seals. Wipe the outside of the jar if it has become greasy. Label, date, and store the jars in a cool, dry location out of direct sunlight.

Raw Pack Bone-Out
For the most part, the raw pack method of canning beef, venison, lamb, pork, chicken, rabbit, or any meat, is identical to the hot pack method. Except you don't precook the meat or cover it with broth. This method is the preferred option for individuals without access to home-grown meat. During the fall months, boneless chicken, beef, pork, and lamb can often be found for a reasonable price in grocery stores. Fall is also the time for deer harvest.

Basically, the raw pack method involves packing raw, boneless meat tightly into a canning jar and then processing it. It's unbelievably simple. This method can be used for all boneless meat. Learn how to can your own meat, and you'll never have to pay outrageous grocery store prices for canned chicken or any other type of meat again.

Here's How to Do It
Cut the preferred meat into jar-sized pieces and pack the pieces tightly into a canning jar.
Add salt if you like:
- ½ teaspoon for Pints
- 1 teaspoon for Quarts

Leave a 1-inch headspace. Wipe the rim of the jar, then apply the lid and band. Next, place the jars in the pressure canner. Close the lid and increase the heat under the canner until steam begins to vent. Allow the canner to vent for 7 to 10 minutes, or as per your canner's manufacturer instructions. Once the canner is vented correctly, apply the weighted gauge to the canner. Bring the canner to full pressure of 10 pounds. Process jars at 10 pounds of pressure (240°F.) for altitudes at 1000 ft. sea level or less. Don't forget you may need an altitude adjustment. See the previous chart.
The processing time for all raw pack bone-out meat and poultry is:
- 75 minutes for Pints
- 90 minutes for Quarts

The Complete Granny Miller

When processing time is finished remove the jars from the canner. Place them on a towel or board well away from drafts. Allow the jars to cool undisturbed for 8 to 12 hrs. After jars have cooled, remove the bands and check the seals. Wipe the outside of the jars if they have become greasy. Label and date the jars and store them away from heat and direct sunlight.

Pick Whichever Method Suits You
The advantage of the raw pack method is that it is a time saver. The advantage of the hot pack method is that there is plenty of broth to work with when you open the jar. Choose whichever method you prefer according to the recipes that you will use and your family's food preferences.

Advice When Canning Venison
Venison can sometimes have a strong flavor that some people object to. The flavor has mostly to do with the gamey, greasy fat that deer, especially bucks, possess. A simple way to avoid the gamey flavor when canning venison is to add a small piece of beef suet about the size of a silver half dollar to the top of the canning jar before it is sealed and processed. The reason this works is that during the processing, all of the fat becomes liquid and migrates to the top of the jar. The suet helps to pull the deer fat away from the meat. When opening a jar of home-canned venison, simply discard the solid wedge of fat on top of the meat.

How to Can Grapefruit or Orange Sections
Home canning grapefruit or orange sections is easy and results in a superior product when compared to commercially canned grapefruit or orange sections. It is a cost-effective way to increase variety in your long-term food storage. Grapefruit, oranges, and other citrus fruits are considered to be high-acid foods. High-acid foods are safely canned by using the water bath method of canning. Grapefruit can be canned alone. But orange sections will taste better if canned with equal parts of grapefruit sections in the jar.

Get Ready
First, gather and assemble the water bath canner, jars, lids, bands, canning funnel, jar lifter, a large kettle, a saucepan, and white cane sugar for making a light syrup.
Begin to heat the water bath canner. Wash the canning jars and bands in hot, soapy water and rinse well. Keep the jars hot. Wash and rinse the grapefruit in warm, soapy water. Rinse well and dry.

Peeling
The grapefruit or oranges need to be peeled. When canning oranges or grapefruit sections, all of the white and fibrous parts of the grapefruit and the seeds must be removed.
Only the 'heart' of the citrus sections should be used. That's because the

white stuff on the grapefruit is bitter and pulpy when canned. When peeling large quantities of citrus fruit for canning, I use a special serrated sandwich knife and a smaller paring knife. Here's a quick way to work through lots of grapefruit.

Think of the grapefruit as a globe. The two flat ends of the grapefruit are the 'North Pole' and the 'South Pole'.

First, thinly slice off the two flat ends of the grapefruit. Next, stand the grapefruit upright on the North or South Pole, and proceed to slice around the entire fruit with the serrated knife until it has been peeled. You'll be slicing on a curve, from the North Pole to the South Pole. When the grapefruit has been peeled, use a small paring knife to free the individual wedges or sections. Once all the sections are removed, squeeze out all of the juice to empty the fruit.

Make Syrup

There are a couple of different ways to fill canning jars with grapefruit. Some recipes call for heating the grapefruit sections in light sugar syrup. Others suggest filling the jars with cold grapefruit sections and then pouring heated light syrup over the sections. Still other recipes use heated orange juice or heated grapefruit juice poured over the grapefruit sections. I usually just pack cold grapefruit sections into a jar and use the juice that was made by squeezing the grapefruit when I peeled it. If I need more liquid, I will pour a small amount of heated light syrup over the sections to achieve a ½ inch head space. A light syrup is made by dissolving 1 ½ cups of cane sugar into 6 cups of water. Heat the sugar and water, stirring until the syrup is hot and all the sugar is dissolved. Set the syrup aside.

Fill the Jars

Add the grapefruit or orange sections to a clean hot canning jar. Pour grapefruit juice,orange juice, or light syrup to within a ½ inch of the jar rim. Slide a non-metallic object down the sides of the jar to remove all air bubbles. Wipe the rim of the jar with a clean, wet cloth and apply a lid and band to the jar. One by one, as the jars are filled, place them aside on a wood board or thick towel. When all the jars have been filled, place them into the boiling water bath canner. Turn up the heat and bring the water in the canner to a full rolling boil. It may take a while for the water to come to a full boil if the fruit was cold-packed. Normally, water in a water bath canner will lose some heat when jars are added. It takes time for the heat to build back up and for the water to begin to boil again.

Process the Jars

Processing time is counted from the time the water in the canner comes to a full rolling boil. Once the water begins to boil, put the lid on the water bath canner and adjust the heat to maintain a gentle boil.

The processing time for both Pints and Quarts of grapefruit or orange sections in a water bath canner is:

The Complete Granny Miller

Processing Time for Grapefruit & Oranges
10 Minutes – Altitudes 1000 ft. or less
15 Minutes – Altitudes 1001 ft. to 6000
20 Minutes – Altitudes Above 6001 ft.

When processing time is complete, remove the jars from the canner. Allow jars to cool undisturbed for 8 to 12 hours. Once the jars are completely cool, remove the bands and check the seals. Wipe the jars clean with a damp cloth; label and store in a dark, cool location.

For planning purposes, 20 pounds of grapefruit will yield about 6 quarts of grapefruit sections. Home-canned citrus will store approximately 12 to 15 months before any noticeable loss of flavor or color. For the best taste, chill canned grapefruit or orange sections at least overnight before opening and serving.

Canning Grape Juice With A Steam Juicer

After 45 years of home canning, I've tried every way known to make grape juice. Trust me, I've put in plenty of time with all the different techniques, methods, and fads for preparing grape juice. Some ways are fun, like putting whole Concord grapes and sugar into a canning jar and then covering the grapes and sugar with boiling water. That method makes a light purple sweetened water and not what I'd call real grape juice.

Other ways are a real pain in the backside. Like boiling mashed-up cooked grapes and allowing them to drip and filter through a sieve and a flannel overnight.

But by far the quickest and easiest way to extract juice from ripe Concord grapes, berries, and other soft fruits is with a steam juicer. A steam juicer is an expensive piece of kitchen equipment, but it is worth every penny.

Steam Juicer
A steam juicer is a 4-piece pan assembly that consists of a lid, a colander pot, a collection pan with a clamped tube, and a water pan. The way that a steam juicer works is that the fruit is first placed into the special covered colander pot. The colander pot sits above the collection pan that rests on a lower pan filled with water. As the water in the lower pan boils, the fruit in the colander pot is softened and ruptured from beneath with steam. The juice begins to flow and drips from the colander pot into the collection pan. When the collection pan is full, the juice is drained and siphoned off with the clamped tube that's attached to the collection pan.

I have found that a whole bushel of Concord grapes will yield approximately 18 to 20 quarts of grape juice when prepared with a steam juicer. It takes about an hour for all the juice to be extracted from a loaded

The Complete Granny Miller

steam juicer. A bushel of grapes is about 4 full loads in a steam juicer. I achieve the best results and extract the most juice by employing a specific trick. I run the grapes in the steam juicer until no more juice is dripping into the collection pan. I then empty the steamed and softened grapes into a large bowl and let the cooked, softened grapes rest while I work on another load of grapes. After I've steamed all the grapes, I take the collected waste grapes from the large bowl and place them back into the colander pot. I then steam the entire heap of waste grapes all at once (they will fit because the bulk has been reduced) for another 30 minutes or so. The softened, waste grapes will usually yield another 1 ½ to 2 quarts of grape juice. After all the juice has been collected, the grape juice is sweetened to taste with pure cane sugar before canning. I use 4 pounds of cane sugar to 18 quarts of grape juice. You may prefer more or less sweetening. The cane sugar readily dissolves in the hot juice. With pure cane sugar, there's never a worry about GM corn syrup or GM beet sugar.

Process the Juice
Fill hot, clean jars with sweetened hot grape juice, allowing ¼ inch of head space. Wipe the rim of the jar and apply a lid and band. Place the jars into a gentle boiling water bath canner, making sure that the water covers the tops of the canning jars by at least 1 to 2 inches. When adding filled jars, the water in the canner will stop boiling. Wait for the water to return to a gentle but steady boil before counting processing time. Process the grape juice according to the following chart.

Processing Time For Grape Juice			
Jar Size	0 ft. – 1000 ft.	1001 ft. – 6000 ft.	Above 6000 ft.
Pints & Quarts	5 minutes	10 minutes	15 minutes
Half Gallon	10 minutes	15 minutes	20 minutes

After the processing time is complete, remove the jars from the canner and allow them to cool undisturbed and free from drafts for 8 to 12 hours. When the jars are cooled, remove the bands and check the seals. Wipe the jars clean, label, and store in a cool dark location.

Pressure Canning Grape Juice
Usually, a boiling water bath canner is used to process grape juice or apple juice. But I pressure can grape juice. There's no food science-tested processing time for fruit juice in a pressure canner that I'm aware of. Maybe it's out there somewhere, and I just haven't seen it. I live at 1250 ft. above sea level and process both pints and quarts of grape juice in a pressure canner at 5 lbs. of pressure for 5 minutes and allow a ½ inch of head space in the jar.

The Complete Granny Miller

How to Can Whole Apple Applesauce

I almost always make canned applesauce over two days. On the first day, the applesauce is actually made. On the second day I can it. My applesauce is a 'whole apple' applesauce. That means the entire apple is used except for the core. Frankly, I don't have the time or patience to peel apples for applesauce and prefer the rosy or caramel color of whole apple applesauce. If you want white applesauce, you'll have to peel the apples to get it. I also routinely pressure can applesauce. Time and cooking fuel are saved by pressure canning applesauce instead of water bath canning it. With pressure canning, the processing time is cut by more than half. For planning purposes, I get about 25 pints of applesauce from 1 bushel of apples.

Some years, my home-grown apples don't look so great. But even though the apples may not be dessert quality, they still make good applesauce or apple cider. Homegrown apples are first gathered and washed in warm, soapy water and thoroughly rinsed. You'd be shocked at how dirty fresh-picked apples can actually be. If you are buying apples from a farm stand or a grocery store, the apples have probably already been washed, and you can skip this step.

After washing, the apples are quartered and cored. Any bad spots are cut away. The apples are placed in a large pot or kettle with just enough water on the bottom of the pot so that they won't stick. About ½ cup to 1 cup of water in a fully loaded 20-quart pot is a good starting point.

The apples are then heated slowly and cooked until they are soft and mushy. Stir the pot occasionally to prevent burning or scorching. It will take a few hours to cook down a half bushel of apples properly. An additional 4 to 5 hours will be required to cool the apples sufficiently, allowing for comfortable handling and proceeding to the next step. Usually, at this point, I stop for the day unless I started early. The cooked apples are put in a cool place overnight, like a refrigerator or on a cold porch. The next day, I pick up where I left off.

Make the Sauce

I use a food mill to puree the cooled apples into a large pot. If you don't have a food mill, you could use an electric blender or food processor, but it will take longer and will be messier. I never sweeten applesauce. But if you prefer a sweet applesauce, this is the point to add sugar to your taste.

The applesauce is next slowly heated until it is hot. Be mindful and stir the pot often because applesauce will scorch or burn. Once the applesauce is hot, ladle it into pint or quart jars, leaving a ½ inch of head space. Slide a non-metallic object down the side of the jar to release any trapped air bubbles. The rim of the jar must be carefully wiped clean. Apply a lid and a band to the jar. The jars are next placed into a pressure canner.

The pressure canner should be vented according to the manufacturer's instructions, usually for 7 to 10 minutes. After the canner has been properly vented, the weight gauge is applied. Processing time for hot pack applesauce in a weighted gauge pressure canner is:

The Complete Granny Miller

Processing Time For Hot Pack Applesauce			
Jar Size	Processing Time	0 -1000 ft.	Above 1001 ft.
Pints	8 Minutes	5 lb.	10 lb.
Quarts	10 Minutes	5 lb.	10 lb.

Processing time is counted from the moment the first jiggle of the weight gauge is heard. After the processing time is complete, the canner is allowed to return to normal pressure. Jars are removed from the canner and allowed to cool away from drafts and left undisturbed for 8 to 12 hours.

After jars are cooled, remove bands and check the seals. Wipe the jars clean, label and date the jars, and store them in a cool dark location.

You can most definitely water bath process applesauce. In fact, that's the way most people do it. The processing time is 20 minutes for both pints and quarts. Processing time is counted from the time the water begins to boil after the jars have been placed in the canner. Applesauce can also be frozen for up to 12 months.

Jams and Jelly -The 4 Things That Make Them

Homemade jams, jellies, preserves, and other jellied fruit products are some of the easiest and most rewarding foods for the beginning home canner. They are fun for the seasoned home canner as well. Homemade jellies, jams, semi-softspreads, and fruit butters add a special touch to any meal and make an appreciated gift. A variety of attractive jars and decorative lids are available for jams and jellies. Often, jelly glasses or jars can be found in thrift stores or at yard sales. Every year, I make different types of jelly and jams to give away as hostess gifts or as last-minute Christmas presents. Basically, there is little difference between the various types of jellied fruit products. They are all made with fruit and sugar, and are jellied to differing degrees of stiffness and consistency. Four essential ingredients are necessary for the production of all homemade jellied fruit products.

They are:
- Fruit
- Pectin
- Acids
- Sugar

Fruit

Fruit gives jellies, jams, butters, and preserves their wonderful flavor and beautiful color. Only the best, top-quality, and barely ripe fruit should ever be used. Fruit furnishes a part of the acid and pectin needed for a successful gel.

Sugar

Sugar is an important ingredient in all homemade jellied food products. Sugar contributes to the gel formation in jellies and also serves as a preserving agent. There are recipes for jams, jellies, fruit butters, and preserves that use other types of sweeteners: honey, fruit juice, corn syrup, maple sugar, and Splenda. In my opinion, the best product is achieved by using only 100% pure cane sugar.

Pectin

Pectin is a naturally occurring substance and is the agent that causes jelly to gel. Most modern jelly, jam, and preserve recipes call for the addition of pectin. Pectin is sold in grocery stores and bulk food stores. There are two types of pectin: powdered and liquid. Keep in mind that they **are not** interchangeable in recipes. Slightly under-ripe fruit contains considerably more naturally occurring pectin than over-ripe or mature fruit. The use of overripe or mature fruit will result in a runny product. This is especially true of jellies. In certain types of fruit, there is enough naturally occurring pectin to make a nice jellied food product without the addition of pectin. Many older marmalade and apple jelly recipes do not use pectin.

Acids

Acids help to add to the flavor of jellied fruit products and help with the gel formation. Like pectin, acid is naturally occurring in fruit. Acid content tends to be higher in just barely ripe fruit and certain types of fruits. Some fruits, like peaches, sweet cherries, or apricots, have little naturally occurring acid and need to have acid added to them to make a dependable jellied product. Fresh lemon juice is the most widely used acid in recipes for semi-soft spreads and jellies.

Making Jam without Pectin

Making fruit jam without pectin is easy. It just takes a little bit more boiling. Jams made without added pectin will be darker and have a more old-fashioned cooked taste when compared to jams made with commercial pectin. The trick to making perfect jams without pectin is a candy thermometer and knowing what sheeting looks like on a metal spoon. When making jam without pectin, you first need to determine the boiling temperature of water in your location on a given day, because the boiling point of water changes with both altitude and atmospheric conditions. To test the temperature of boiling water, you'll need a jelly or candy thermometer. Once you know what temperature water boils at, simply add

The Complete Granny Miller

9°F to that number for perfect jam every time.

Peach Jam without Added Pectin

- 4 ½ cups of peeled, pitted, and crushed ripe peaches
- ¼ cup of fresh lemon juice
- 7 cups of white cane sugar

Measure out crushed peaches and place them in a large kettle. Flat-bottom jam kettles are perfect for this. But any good heavy 8-quart pot will do. Add the lemon juice and sugar to the peaches and stir well. Place the jam kettle on high heat and stir constantly until the mixture comes to a full rolling boil. A full rolling boil is a boil that cannot be stirred down.

Once the jam mixture has begun to boil, occasionally test the mixture for correct temperature and sheeting. To take the temperature of cooking jam, place the candy or jelly thermometer into the center of the boiling mixture. But take care that you don't rest the thermometer on the bottom of the pan. You want the temperature of the jam, not the pan. Remember, you need a temperature that is 9°F above the boiling point of water.

'Sheeting' on a spoon is another method to double-check and test jam or jelly. Sheeting is tested by dipping a cool, clean metal spoon into the mixture and quickly lifting it up and to the side. You are looking to see two perpendicular drops of jam that will run together to form one thick drop on the edge of the spoon. The jam mixture will form a jelly sheet on the spoon. The characteristic layer of jam on a metal spoon is called sheeting. It's the method that our great-grandmothers used when they tested for the correct jelly or jam temperature. Keep in mind that jams and jellies will thicken as long as they are heated, and it's easy to overcook jam if you're not careful.

There is another and more modern method for testing jam. The test is performed by cooling a small amount of hot jam on a plate and placing the plate in the freezer compartment of a refrigerator. While conducting the test, you'll need to remove the kettle from the heat so that the jam doesn't accidentally overcook.

To conduct the test, place a small amount of hot jam on a clean plate and then put the plate in the freezer for a few minutes. If the jam forms a gel, it is probably done. But if the jam is still too runny, it needs more time on the stove. After the jam is cooked, and you are confident that it's the right consistency, remove it from the heat. Set it aside for about 5 minutes to allow any foam to collect on the top.

Now carefully remove and skim off as much foam from the top of the jam as you can with a slotted metal spoon. It helps to rinse and clean the spoon between skimmings. The foam does not harm the jam. It's simply removed because of its appearance. Foam tends to migrate to the top of a sealed jar of jam or jelly and has a rubbery look and feel to it. You won't win any blue ribbons at the local county fair with foamy jam.

After the foam has been removed, pour the jam into hot ½ pint jars, leaving about ¼ inch of head space. Wipe the jar rims clean and seal the jars with a

The Complete Granny Miller

lid and band. Process the jars for 10 minutes in a gentle water bath. Processing time is counted from the time the water begins to boil. When processing time is complete, remove the jars from the canner and place them on a wooden board or a thick towel. Allow the jars to cool undisturbed for 8 to12 hours. When completely cool, check the seals and remove the bands. Store the jam in a cool dark location.

Jam Making Tips
- Make only enough jam for one year. Jams and jellies lose quality if stored for too long.
- Floating fruit is reduced considerably by stirring the jam mixture after the foam is removed and the jam has cooled down a bit.

Canned or frozen fruit may be used when making jams and jellies. In fact, a superior strawberry jam is made from frozen strawberries instead of fresh ones. I think the best pineapple jam comes from crushed canned pineapple. Modern canning lid systems work better for jelly than a layer of paraffin in a jelly glass.

Pumpkin Butter

Pumpkin butter made from fresh pumpkins is delicious. It doesn't taste the same, nor does it have the unappealing texture that pumpkin butter made from commercial pumpkin has. It's an extra effort to cook and prepare fresh pumpkin rather than opening a commercial can of packed pumpkin. But I

think that the finished product is more than worth the effort.
If you want to make pumpkin butter from fresh pumpkins, it is best done in two parts.
In the first part of the process, you must cook the pumpkin and puree it.
The second part of the process is the actual cooking of the butter.
There are a few different ways that you can cook and peel a pumpkin. I like to cook pumpkin in a pressure cooker. It's the fastest and easiest way I know of to get the job done. Some people use a microwave oven to peel pumpkin, but boiling the pumpkin will also work. So will baking it in a conventional oven until it is soft. The goal is to cook the pumpkin, remove the rind, and then puree it. Use whatever method works best for you. With that said, here's a way to make pumpkin butter:
Gather and collect fresh pumpkins from the garden. I think the smaller pie-type pumpkins taste the best.
Wash the pumpkins well. Cut them in half and scoop out the seeds with a large spoon or ice cream scoop. Cut the pumpkin into pieces small enough to fit inside a pressure cooker. Cook the pumpkin pieces in a pressure cooker for 9 minutes at 15 pounds of pressure. Quickly cool the pressure cooker by placing it in a sink and running cold water over it. The pressure

The Complete Granny Miller

cooker method is quick and cooks the pumpkin to perfection. Pressure cooking also makes the rind peel off easily.

After the pumpkin is cooked and peeled, place it into a large bowl to mash it. Once it is mashed, run the pumpkin through a food mill to puree it. Some cooks prefer to use a food processor or a blender to puree pumpkin because it makes for a smoother and creamier texture. But I use a food mill because I prefer a more old-time textured product. Once the pumpkin is pureed, it's time to move on to the second step.

In a large pot, combine:
- 3 cups of pumpkin puree
- 2 cups of white cane sugar
- 3 tablespoons fresh lemon juice
- 1 tablespoon ground cinnamon
- ½ teaspoon ground cloves
- ½ teaspoon ground ginger
- A dash of ground red pepper for bite

Depending on the thickness of the mixture, cook the pumpkin butter over low to medium heat for about 30 to 45 minutes, or until it rounds up nicely on a spoon. Stir it occasionally to prevent it from sticking or scorching. When the butter looks and tastes ready, ladle it into clean, hot ½ pint jars, leaving a ¼ inch head space. Apply a band and lid to the jar and process the jars for 15 minutes in a boiling water bath canner. After 15 minutes, remove the jars and allow them to cool on a towel undisturbed for 8 to 12 hours. When the jars are completely cool, remove the band and check the seal. Most often, I triple or quadruple the above recipe. Ordinarily, I get about 4 half pints of pumpkin butter for every 3 cups of pumpkin that I use.
And just so you know, since the mid-1990s, there has been a controversy regarding the safety of home-canning pumpkin butter. At present, home canning pumpkin butter is not recommended by so-called canning authorities and experts. As far as I can tell, the safety issue is a theoretical one. That's because pumpkin is a low-acid food. The only safe way to can low-acid foods is with a pressure canner. The difference with pumpkin butter is that the addition of a large amount of sugar and lemon juice radically alters the pH. The heavy sugar load and acidic lemon juice make the pumpkin butter safe to process by the boiling water bath method. At least that's what everybody thought and did before somebody wrote a food science paper and freaked out home canners everywhere.
Apparently, there is a possibility that you could get botulism from pumpkin butter. So, these days, experts are advising home canners to either freeze their pumpkin butter or store it in the refrigerator. But frankly, I don't have much use for warnings from food science experts that are 35 years too late. You do what you think is best with your pumpkin butter. But as for me, I'm hopelessly stuck in my ways and haven't died yet from poisoned pumpkin butter.

The Complete Granny Miller

Storing Home-Canned Foods

For the best quality home home-canned food should be stored in a cool, dry location. Storage temperature should ideally be between 50°F to 70°F. Canned food needs to be protected from excessive heat, freezing, and dampness. Heat causes home-canned food to lose quality rapidly.

All home-canned food should be stored well away from direct sunlight, hot pipes, heat ducts, gas or electric ranges, and wood heat appliances. Do not store canned food in an uninsulated attic. Freezing cold temperatures do not cause canned food to lose quality. But it may damage the seal on a jar. Remember, once a seal is damaged, air enters a jar and spoilage begins. Also, repeated freezing and thawing may soften food or make it mushy. If home-canned food has to be stored in an unheated area, jars can be placed in heavy boxes and covered with a wool or Mylar blanket, or a heavy layer of newspaper.

Canned food should never be stored in a damp area. Dampness may corrode the metal jar lids and compromise the seal.

In the last 10 or 15 years, there has been an increased interest in preserving food and storing food without the use of electricity. Non-electric food storage was, at one time, basic household information that every housewife or farmwife knew. What follows next are three different types of perishable foods that can all be stored in crocks.

How to Preserve Eggs with Water Glass

Did you know that there are a few different old-time ways to preserve fresh eggs for months without electricity? It's true.

Freshly laid eggs can be successfully preserved by being kept in a water glass solution; in a lime water solution; or by being coated with mineral oil, Vaseline, paraffin wax; or by being buried in sawdust, sand, oats, and salt.

Of all the old-time methods of preserving fresh eggs, the water glass method gives the best and most dependable result. Fresh, unwashed eggs kept in a solution of water glass will remain good and useable for 4 to 5 months when properly collected and stored.

Water glass or 'liquid glass' is sodium silicate and is the generic name for sodium metasilicate (Na_2SiO_3).

Nowadays, water glass has become hard to find. At one time, it was readily available in drug stores, hardware stores, and building supply warehouses. Water glass is alkaline in nature and has the taste of washing soda. It is used for general cleaning purposes, to seal unfinished cement floors, and as an adhesive. Water glass is a clear, slightly syrupy liquid that comes already dissolved in a gallon bucket. It has been within my lifetime that preserving eggs in water glass has fallen out of favor. Methods of egg storage have changed due to almost universal household refrigeration and factory farms.

The Complete Granny Miller

Today, eggs are cheap and readily available. Fresh eggs are no longer hard to come by during the winter months.

If you don't already know, hens will naturally cease egg production once daylight hours are decreased during the winter. Many people who have backyard chickens and electricity will put a light in the hen house during the winter to force hens to lay. Lighted hen houses, hens confined in small battery cages, and long-term cold storage are the reasons that there are cheap eggs in grocery stores during the winter months. Without electric lights, most hens will only lay hit or miss during the winter.

Before refrigeration became commonly available, keeping fresh eggs in a crock of water glass was the preferred method of egg preservation. For many American families prior to the Rural Electrification Act of 1936, water glass was the only way that they could manage to have eggs during the winter months, when hens are naturally discouraged from laying due to cold, dark days. By saving surplus eggs during the spring and summer when eggs are plentiful, farm families were guaranteed a steady supply of eggs through the winter months. Sadly, within the course of two generations, what was once everyday household knowledge has been lost and forgotten. Water glass has gone the way of pants and curtain stretchers, and wire bail canning jars. To preserve eggs in a solution of water glass, you must first obtain the water glass, which sometimes is easier said than done. Hardware stores and big box building centers seldom keep it in stock, but it can be found online.

How Water Glass Works

Eggshells are porous. That's why an incubating chick embryo can breathe. Eggs will spoil and lose freshness due to bacteria passing through the shell, and by the evaporation of moisture from the interior of the egg. The way that water glass preservation works is uncomplicated and straightforward. The water glass blocks and fills the pores of the eggshell, thereby preventing bacteria from entering inside the egg and moisture from leaving the egg.

Eggs for Water Glass Preservation

Eggs that are to be used for water glass preservation must be completely fresh and clean. They must not have ever been washed. By washing a fresh-laid egg, the protective coating that is natural to the egg is removed. It is acceptable to lightly wipe an egg with a dry cloth if it is a little soiled. The best eggs are collected from fresh, clean nest boxes and will have no cracks or imperfections. One cracked egg will spoil an entire crock of eggs. Do not use washed grocery store eggs for water glass storage. They absolutely will not keep.

It has often been said that the best eggs for water glass are collected during the spring months of March, April, and May. I think the reason for this is that the weather has not turned too hot. The cooler weather keeps an uncollected egg fresher in the nest box. That said, I collect eggs for water glass during June, July, and August for storage. Eggs are plentiful then, and we enjoy cool summers in Western Pennsylvania.

Old timers would not permit a rooster to run with the hens for up to a month

The Complete Granny Miller

before eggs were collected for water glass storage. It was from concern that a fertile egg could be beginning to develop into a chick embryo. I don't think this is a genuine concern as long as the eggs are collected daily and stored correctly. But no matter what I think. The practice of early 20th-century housewives and farmwives was always to crack eggs that had been stored by any method into a separate bowl for examination before cooking with them. After all, who am I to argue with time-honored kitchen tradition and experience?

Prepare the Water

Water glass always needs to be diluted with water. The recipe I use is an 11-to-1 ratio recipe. Or 11 parts of water to 1 part water glass (sodium silicate); or 11 quarts of water to 1-quart water glass; or 11 pints of water to 1 pint of water glass. You get the idea.

It works out to 1 quart of water to 1/3 cup of water glass. Just so you know, some recipes will give a 9 to 1 or 10 to 1 ratio. I have no experience with them.

The water for dilution should be measured out, boiled, and then allowed to cool completely. Many older recipes recommend rainwater.

Lay Eggs Down In the Crock

Sterilize a clean ceramic crock, plastic bucket, wooden keg, or other container with boiling water. You want to destroy any possible yeast, enzymes, or bacteria that may be present. Almost any water-tight container will work. But metal should be avoided. Pour the cooled water into the crock and then add the water glass. Stir well. The water must be completely cool. You don't want the hot water to cook the egg. Place the fresh eggs pointed side down into the crock. You can fit many eggs into a crock, and eggs can be stacked on top of one another until the crock is filled. Make sure that at least 2 to 3 inches of water glass solution covers the eggs at all times. The crock should be tightly covered.

The best egg storage success is obtained when the crock is stored in a cool, dry location. A fresh, clean root cellar, spring house, garage, or cold basement storage area is ideal. Clean, fresh eggs can be added daily as the season progresses. If water ever needs to be added to the crock, just make sure that it has been boiled first.

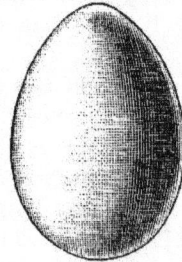

At a later date, when the eggs are needed for cooking, simply remove the eggs from the crock and wash them. Break the eggs into a separate bowl to check them by smell and visual examination. Eggs that have been stored with water glass may break when they are boiled, so use caution if you intend to cook them by that method.

Fresh collected eggs will store well for between 3 to 5 months without too much loss of quality. The viscosity of the egg white will have changed, but

The Complete Granny Miller

the flavor is still good and acceptable for general cooking and baking purposes. Sometimes the yolk will take on a very dark orange-red color, but it is harmless.

How to Store Meat in a Crock

There is an old method of storing some types of meat and fish known as 'potting' or 'crocking'. It is a non-electric method of meat and fish storage that was widely prevalent before the advent of home canning. Potting or crocking meat and fish is a method that the USDA no longer recommends because of the potential for botulism food poisoning. But in many areas of the world, especially in France, it is a method that is still widely practiced. In fact, up until the Second World War, many American housewives and farmwives used the crocking or potting of meat for short-term food storage. After World War II, the USDA made a concerted effort to educate American housewives and to improve the safety of home food preservation. It was during this period that the USDA began to strongly emphasize pressure canning as the only recommended safe method of preserving low-acid foods. But old habits can die hard. Many rural American households refused to give up the old ways and continued with food preservation the way they had always done before.

What Exactly Is Potting or Crocking?
Potted or crocked meat is meat that has usually been thoroughly cooked and is laid down in a sterile crock. It is then encased with lard, butter, or some type of grease. The crock is covered and is usually stored in a cool, dry location. The science behind crocking or potting is, in theory, that cooking destroys potentially harmful bacteria in the meat. The subsequent encasement of the meat by fat seals the meat from the air, and no further spoilage can occur.

Potting, in theory, is similar to home canning. That is, heat destroys bacteria, and a lack of air or a vacuum leaves the food in a type of suspended animation. The risk with potting or crocking is that the fat or grease in the crock can insulate botulism spores that may not have been destroyed by cooking. Botulism spores in meat could be protected in the anaerobic environment of a grease-laden crock. This is the exact same type of risk associated with home canning bacon or butter, which has been circulating on the Internet.
But here's where the talk about risk and the USDA guidelines gets tricky. The actual risk versus the theoretical risk of home canning or potted bacon

is unknown. Without expensive laboratory testing, and because of the wide range of individual kitchen practices and sanitation, it is impossible to know what exactly is going on in any particular crock or Mason jar.

Botulism is an old and dangerous food safety issue. The word botulism derives from the Latin word 'botulus' or sausage. Food poisoning and bad sausages have been around for a long time. While my crocks of sausage patties or bacon may be safe, my neighbor's crocks may not be safe. With that said, what follows is to be used at your own risk.

How to Crock Meat

Bacon, sausage patties, sausage links, and pork chops can all be stored in a crock covered with grease. I have no personal experience with any other types of meat or fish.

Wash a ceramic crock with hot, soapy water. Then, sterilize the crock by pouring boiling water into it. Hold the boiling water in the crock until just before filling it with meat.

Next, cook the sausage patties, sausage links, or bacon thoroughly. You want the internal temperature of the sausage or bacon to reach well above 250°F. After the meat is fully cooked, remove it from the heat and let it sit in the grease.

Empty the water from the crock and wipe the crock dry with a clean towel. Place hot grease in the bottom of the crock so that the bottom is completely covered. The grease should be about 1 inch deep. Next, place a layer of cooked sausage patties, sausage links, or bacon into the crock. Cover with hot grease.

Add another layer of meat and repeat, adding hot grease after each layer. When the crock is full or you run out of meat, cover the meat with at least 2 to 3 inches of hot grease. Cover the crock with a plate or a cloth. Store the crock in a cool, dry place.

When you want to eat the sausage or bacon, remove the meat from the grease carefully with a fork. Place the bacon, pork chops, sausage, or whatever you have, in a frying pan and re-fry. Heat the meat thoroughly. The internal temperature of the meat must reach at least 250°F once again. With bacon, this takes about 3 or 4 minutes on high heat. With sausage patties or links, it takes about 6 or 7 minutes. The meat must be reheated to 250°F to kill any potential botulinum toxin.

There is absolutely no taste difference in sausage or bacon when stored by this method. Crocked sausage patties and links are superior in flavor and taste to canned sausage patties or links.

Why I Don't Can Bacon

One reason I'm not a fan of canned bacon is that, from my perspective, it is a waste of time, stove fuel, Mason jars, lids, bands, paper, and human effort. Why go through all the trouble to fiddle with a pressure canner, paper, scissors, canning jars, lids, and bands, not to mention the wood, gas, coal, or electric power needed to fire the pressure canner, when you can have a

The Complete Granny Miller

product that takes lots less time and is just as unsafe? In fact, a good argument could be made that crocking or potting bacon is actually safer and less risky than the popular Internet method of canning paper-wrapped bacon. Here's why.

A popular Internet method for canning bacon involves wrapping bacon in layers of rolled paper. It is a copycat of expensive canned bacon manufactured by large industrial food companies. Paper-layered bacon is inserted into a quart-size canning jar and processed at the standard canning pressure and processing time of 10 lbs. of pressure for 90 minutes at altitudes of 1000 ft. sea level or less. The problems I see with this method are at least twofold.

First, because this method is untested in a food science laboratory, there is no way to know for sure that the core temperature of the jar has reached 240°F for a long enough time to guarantee 100% bacteria kill. With the addition of layers of paper, there is an added layer of insulation. The extra layers of rolled paper surrounding the bacon may act as a cushion and insulate the inner core of the jar from the heat. The fat present in the bacon, in theory, can insulate the spores of Clostridium *botulinum*. The presence of heat may actually begin to activate dormant spores, which will soon be in an anaerobic, moist environment (an air-tight canning jar), and begin to produce toxins. Secondly, paper is not inert. Especially when heated. Unless food-grade paper is used, toxic and unwanted chemicals can be leached into the bacon.

How to Store Fresh Lemons in a Crock

For long-term storage, fresh lemons keep best in a tightly sealed plastic bag in the refrigerator. But if you don't have a refrigerator or if space in the fridge is limited, there is a time-tested low-tech way to store fresh lemons. Fresh lemons will store well for a couple of months by being completely submerged in a ceramic crock of cold water.

Simply place fresh lemons in a clean ceramic crock and cover them entirely with cold water. Set a small plate or water water-filled plastic bag on top of the lemons to keep them submerged and prevent them from floating. Allow the water to cover the plate.

If you use a water-filled plastic bag, add more water to weigh the bag down if necessary to keep the lemons submerged in the cold water. Store the crock in a cool location like an unheated basement, garage, or root cellar. Change the water every week and you'll have a supply of fresh, juicy lemons whenever you want. This method also works well for cranberries.

How to Render Lard and Tallow

Pure lard without preservatives can be hard to come by. And tallow is almost impossible to find anywhere and at any price. If you don't already know, lard comes from pigs and tallow comes from cows. Both products are made from the fat of those animals. I use lard in some baked goods and tallow for soap. Lard makes a flaky pie crust, and tallow makes a beautiful and creamy white hard soap.

Rendering both tallow and lard is simple. The rendering is essentially the process of cooking down pork fat (lard) and beef suet (tallow) until all the fat has been melted away from the gristle or tissue. The fat becomes liquid and is strained into a container. The fat is then allowed to cool and harden, resulting in a finished product known as lard or tallow.

The best results for both lard and tallow begin with the proper selection of fat from home butchering or a butcher shop. For lard, I always ask for the 'leaf lard'. Leaf lard is the fat that surrounds the kidneys of a pig. It is a rich, creamy white, clean fat. But any type of pork fat will do. For tallow, I prefer the big chunks of suet from around the internal organs of a cow. The rendering process goes much faster if the fat is cut into small pieces before it is heated. Time with this step can be saved by having the butcher grind the fat using an electric grinder, rather than cutting or grinding it by hand at home.

Once the fat has been cut or ground into small pieces, it is placed in a pot with about a ¼ inch of water in it. The water is added to the pot or kettle to prevent the fat from burning. The pot is then heated slowly until the fat starts to melt. I try to do my rendering outdoors because the smell of the melting lard and suet can be unpleasantly strong. Also, rendering indoors can leave a layer of grease on the stove, walls, and cupboards. An outdoor LP camp stove works well for this.

As the fat starts to melt, it needs to be stirred occasionally. Suet will begin to turn soft and mushy as it is cooked down. The smell is strong: like a cheap prime rib dinner. Lard acts and looks different from the suet while it is being rendered. The lard seems to fry as the cracklings (bits of pork) start to fry out and to settle on the bottom of the pot.

After the lard pieces have released as much fat as they are going to, the crackling pieces are scooped out of the pot with a slotted spoon. The cracklings can be salted and eaten. The clear rendered lard is then strained and poured into containers. Lard will keep in the refrigerator or freezer for well over 6 months. If it is canned, it will keep for almost indefinite periods. Canned lard is good for future soap making. But it may not be so good for

The Complete Granny Miller

food purposes. Be aware that lard is a low-acid food. Theoretically, canned lard may carry with it the risk of botulism due to the fat possibly insulating botulism spores. To can the lard, simply pour it into a clean hot canning jar, and apply a lid and band. Once the jars have thoroughly cooled, remove the band and check the seal. ***Can lard at your own risk.*** I personally never can lard for food purposes. I store it in the refrigerator or freezer.

Straining and finishing the tallow is not as straightforward as straining the lard. That's because the tallow or suet doesn't completely liquefy like lard does. It won't just pour into a container. The tallow needs to be transferred into another container so that it can harden. I use a large stainless-steel bowl for that purpose. Softened beef fat will have a semi-jelly-like consistency. It needs coaxing through a food mill, strainer, or sieve. I have often used a fine mesh strainer and a spatula to work the hot beef fat into a pan or bowl. But in recent years, I have elected to use my food mill with an excellent result.

Once the tallow is strained, put it into a pan or bowl with about 1 inch of water in the bottom. Place it in the refrigerator overnight so that it will harden. The next morning, the hardened tallow is easily removed. It is now ready to be used in soap, black powder guns, candles, bird feeders, or whatever else you need tallow for. I keep my tallow in the freezer. But plenty of people just keep it on the pantry shelf or in the refrigerator. Well rendered tallow will last a long, long time.

How to Restore Hard Honey

Raw, pure, unfiltered honey can go hard or crystallize in storage. But it's an easy fix. To return crystallized honey to its liquid form, simply heat it gently. But beware. If you heat honey at too high a temperature, you will change the color and the flavor of the honey. Heating honey multiple times will reduce its quality.

I store honey in a Mason jars. When I find a jar that has gone hard, I simply heat the jar slowly in a pan of water. Once the honey returns to its liquid state, I pour it into another clean Mason jar using a jar funnel.

All pure, raw honey will, over time, crystallize. The reason store-bought honey doesn't readily crystallize is that it was heated to a high temperature before it was bottled. Honey naturally crystallizes at room temperature because the glucose molecules separate from the water. When this separation occurs, the glucose begins to attach to seed crystals in the honey. Once crystals are formed, the glucose and some of the other sugars in the honey are attracted to the crystals, and rampant crystal growth begins. The speed at which raw honey will harden or crystallize depends upon the source of nectar, the temperature at which the honey is kept, and the seed crystals in the honey. If honey is well sealed, it has an indefinite shelf life. In fact, 3,000-year-old honey good enough to eat has been found in Egyptian pyramids. Honey has marvelous antiseptic and soothing properties. Raw honey mixed with fresh lemon juice will help ease a sore throat. And honey mixed with whiskey helps to calm a dry cough.

The Complete Granny Miller

The recipe that follows was the most widespread crockpot yogurt recipe on the Internet for years. I had no idea when I wrote it how popular it would become.

Or, for that matter, that the advertising revenues generated from it would pay most of the hosting service fees for my old website, GRANNY MILLER.

How to Make Foolproof Crock Pot Yogurt

Homemade yogurt is actually simple to make. And if you ask me, the easiest and most dependable way to make yogurt at home is with an electric crock pot. All you need is a crock pot, a thermometer, fresh milk, and a little bit of the proper type of bacteria.

There are many recipes for homemade yogurt. Some of them actually make what I would consider to be yogurt. But plenty of them don't. Homemade yogurt does not have the same firm consistency as store-bought yogurt. Store-bought yogurt is a fake-out. It is set and stiffened with pectin, milk solids, and other thickeners. Yogurt created without added pectin or powdered milk will have a top layer of whey that makes the yogurt thinner. Whey is the natural liquid by-product of cheese and yogurt making. The main thing to remember about homemade yogurt is that if you want to make it thicker, the whey needs to be strained from the yogurt. The more whey that is removed from the yogurt, the firmer the final product. In fact, if you strain off most of the whey from yogurt, you'll end up with a delicious soft cheese known as yogurt cheese.

The crock pot method of making yogurt produces dependable results. It's pretty much foolproof as long as you follow the directions faithfully and precisely. If you want success with the following recipe, don't improvise.

Here's What You'll Need
- 1 gallon of milk. 4 quarts of any type of milk from a mammal will work
- 2 tablespoons of yogurt. The bacteria needed for making yogurt must come from somewhere. If you don't already have a starter, you'll need to use some yogurt. It can be any kind of yogurt, but it must contain both live and active cultures of Lactobacillus bulgaricus and Streptococcus thermophilus. Check the label to ensure you have the correct starter bacteria.
- an electric crock pot
- some type of cooking or floating dairy thermometer
- a whisk or fork
- a colander
- muslin, Plyban cheesecloth, or some type of loose-woven cloth

- bath towel or woolen scarf
- an oven or other draft-free warm location

Here's How to Do It

Place the gallon of milk into the crock pot and cover it. Slowly heat the milk until it reaches a temperature between 180°F and 190°F. It is essential to heat the milk to at least 180°F. The purpose is to make the milk sterile and free from all bacteria. The only bacteria you want in the milk are Lactobacillus bulgaricus and Streptococcus thermophilus, which you will intentionally add when you inoculate the milk. This step is crucial with raw milk. You must eliminate other bacteria to get consistent and reliable results when making yogurt. Competing bacteria can cause problems.

Allow the milk to cool naturally and undisturbed to a temperature of 110°F. It takes about 3 ½ to 4 hours to cool the milk down to that temperature. It is critical to the success of the yogurt that you catch the milk at 110°F. This is the ideal temperature for inoculating yogurt. A milk temperature any higher may kill the added bacteria. And if the temperature is too cool, the bacteria will not thrive.

If you are using non-homogenized or raw milk, a wrinkly skin will form on the top of the milk as it cools. The skin should be carefully and completely removed. If you don't remove all of the milk skin, you'll get nasty, hard flakes in your yogurt.

Pour about 1 cup of warm milk into a separate cup or small bowl. Add 2 tablespoons of starter yogurt to the 1 cup of milk. Do not add any extra yogurt; two tablespoons of starter are all you need. The bacteria need ample space to grow and won't develop properly if crowded.

With a fork or whisk, gently but thoroughly stir the starter yogurt into the cup of milk to inoculate it. Next, slowly pour the inoculated milk back into the crock pot and stir it gently by moving slowly from side to side. Do not stir in circles. Use a careful and slow up-and-down lifting motion moving across the length of the crock. After the milk has been inoculated, carefully lift the covered crock out of the electric base and place it into a cool oven. Wrap a bath towel or woolen shawl snugly around the crock and leave it undisturbed overnight or for about 10 to 12 hours. You want the milk to stay nice and warm. An oven with a pilot light or electric light turned on works great. Do not disturb the milk and keep the oven door closed. If you open the oven door, you may have a yogurt failure.

After 10 or 12 hours, your yogurt should be solid with a layer of whey on top. If you prefer a thicker yogurt, you'll need to drain or carefully pour off the whey. I do this by pouring the yogurt into a colander lined with Plyban cheesecloth, which has been placed on top of a large pot. A rectangular piece of muslin or a clean dish towel can also be used. Don't use regular cheesecloth, as the weave is too loose and the yogurt will slip through.

As the whey drains away from the yogurt, it is collected into the pot. Whey can be used later for another food purpose or fed to chickens, pigs, or another animal. It takes about 2 hours of draining to make a thick natural yogurt, and about 3 or 4 hours to make Greek-style yogurt.

Once the yogurt has thickened to preference, lift the cheesecloth from the

125

colander and carefully dump the yogurt into a covered dish or large container. Store yogurt in a refrigerator to keep it sweet-tasting. Some people prefer a tart yogurt and leave it out at room temperature for over 24 hours. The longer yogurt stays at room temperature, the more tart it will become. Remember to save a little bit of yogurt back, and you'll always have starter for the next batch.

Food Storage

At one time, every rural household stocked up on enough basic food and supplies to see them through the winter months. When food was plentiful, it was plain old agrarian common sense to store food for lean times.

Even the most cursory observation of the natural world teaches us that there is a built-in food cycle for all creatures. Summer is the time for abundance, and winter is the time of scarcity. It takes no great intelligence or cleverness to acknowledge changing seasons or to provide or plan for an uncertain future. Squirrels and insects do it every fall without fail.

But with the development of suburban lifestyles and the arrival of large grocery stores, most people don't see a need for creating secure home stores of food and supplies. Modern people, for the most part, are disconnected from the source of their everyday food supply and basic needs. Many assume that no matter what happens, somehow food will always be available to them. It's a dangerous and foolish assumption.

What follows is a food and supply list that was first published on GRANNY MILLER in the early summer of 2013. The list was popular with readers for a couple of different reasons. First, I think people were curious to see the amount of food and supplies two middle-aged adults went through in a year. Secondly, I believe readers were seeking concrete direction and inspiration to plan their stores of household food and supplies.

A Year's Worth of Food and Supplies

Every June, I start anew the annual cycle of putting aside a year's worth of food and supplies. The goal is to have all the food and supplies I need in my pantry, cupboards, closets, and cellar by Halloween.

My practical food storage education first began in the late 1980s when I attempted to store an entire winter's worth of food and household sundries. I live on a back road in the Snow Belt. Thirty years ago, it was not at all unusual for us to be snowed in a few times every winter. Often, we were housebound for up to a week or 10 days at a time.

My first winter food storage efforts proved to be such an incredible convenience and time and money saver that by 1990, I extended my pantry and household stores to last one entire year. I took a lesson from my

The Complete Granny Miller

husband's Appalachian grandmother, and coincided the beginning of my food storage with each year's new gardening and growing season.

 My household food year begins with planting a garden in spring and ends with the final harvest in the fall. What follows below is this year's (2013 - 2014) pantry and supplies list. It's the master supply list that I will carry in my purse and use for the next five months to help me store or buy a year's worth of food and supplies for two older adults. In practice, the food and supplies last longer than a year. I almost never run out of anything and always end the year with a surplus of goods to be rolled over into the following year.

Food and supply lists are fun to look at. But this list probably won't benefit anyone except my husband and me. That's because no two households have the same foodways, food preferences, living arrangements, or dietary needs or restrictions. If you study the list below, you'll notice that there are no tree nuts or chocolate listed. That's because chocolate provokes migraine headaches in both my husband and me, and tree nuts are a migraine trigger for me. Chocolate chip cookies, hot chocolate, chocolate cake, brownies, and chocolate pudding don't happen at my house anymore. You should keep that in mind as you peruse the list.

You also won't find any margarine, orange juice, turkey, ready-to-eat breakfast cereals, sweetened drinks, or pasta because we don't eat those things. Nor will you find eggs, fresh fluid milk, or beer. We keep chickens, get milk from our dairy cows, and buy beer by the case when we run out. You won't find dog or cat food either.

What you will find is lots of canned pink salmon, tomato juice, and white vinegar. That's because I eat lots of salmon, like tomato juice for breakfast, and clean my house with white vinegar. You might not clean house or guzzle tomato juice at breakfast the way I do. That's why it's essential to understand what is happening in your own home and what you actually eat and use. Because every household is unique, I recommend that anyone serious about the storage of food and supplies do a yearly household inventory.

The items listed below that are marked with an asterisk * are items that I will buy at the grocery store or elsewhere. They are things that either I can't produce myself or don't care to. Keep in mind that my home garden and orchard are the mechanism by which all food storage and production in my home depends upon. Without a garden, I'm unsure how to manage my kitchen and pantry economically, or where to source food and how to afford it. The list that follows does not take into account fresh fruits and vegetables that are consumed from my home garden and orchard. In my location, most of the fresh food production and utilization occur between May and October. The fresh food year starts with asparagus and lettuce in May, and ends with apples in October, and Brussels sprouts sometime around the beginning of December. Literally hundreds of pounds of food are consumed and processed for storage during those six months. Don't forget that when

The Complete Granny Miller

you read through my list, I have quite a bit of food and some supplies leftover from last year. You are not looking at a complete food or supply list for one year. To get a better idea of what one year's worth of food and supplies looks like for two older adults, you must add what I already have in my pantry and cupboards.

So, without further delay, here's my 2013 - 2014 Pantry & Supplies List

Food List
Meat
6 whole chickens, *# 25 chicken thighs, * # 10 chicken breasts, ½ pig, # 20 fresh sausage, *#20 bacon, #*10 kielbasa sausage, *5 pkgs. hot dogs, *60 cans of pink salmon, *10 cans smoked kippers, # 10 beef suet, # 10 leaf lard,
(1 Dexter steer to be processed in December - approx. # 250 beef)

Vegetables from the Garden
56 quarts tomato juice, 63 quarts tomatoes, 18 pints red beets, 9 pints carrots, 8 quarts frozen tomatoes, 20 quarts frozen broccoli, 5 quarts frozen zucchini, 5 quarts frozen yellow summer squash, 8 quarts chopped frozen green peppers, 8 quarts chopped frozen onions, 10 quarts frozen sweet corn, 10 quarts frozen acorn squash, 5 quarts mashed frozen pumpkin, 6 quarts frozen cabbage, 5 quarts frozen Brussels sprouts

Fruits
63 pints apple sauce, 21 quarts of peaches, 14 quarts of pears, 42 quarts grape juice, 6 pints cranberry juice, 25 pints frozen blueberries; blackberries & raspberries as God provides to go to freezer, *4 cans sliced pineapple, *# 8 box of raisins, *5 boxes prunes, * 2 bags of cranberries to freezer

Fresh Fruits & Vegetables to Cellar
potatoes, apples, pears, onions, garlic, 2 pumpkins, carrots, cabbage, squash, tomatoes, apple cider

Frozen Food from the Grocery Store
*10 boxes frozen spinach, *6 bags frozen peas, *5 bags lima beans, *4 bags mixed vegetables, *6 boxes breaded fish, *6 boxes Pirogies, *10 frozen pizzas

Dairy
*# 30 butter, *15 pkgs. cream cheese, *30 cans evaporated milk

Condiments, Sauces & Jellies
24 - 36 ½ pints of jam & jelly, 14 quarts apple butter, 12 pints salsa, * 3 jars sweet relish, * 1 bottle ketchup, * 2 squeeze bottles yellow mustard, * 3 bottles barbecue sauce

---

The Complete Granny Miller

Canned Soups & Stews
14 quarts lentil soup, 14 quarts canned beef chili, 14 quarts
lamb stew, 14 quarts beef stew, 14 quarts beef vegetable soup,
7 quarts split pea soup, 18 pints chicken cubes

Pantry Backbone
3 quarts red wine vinegar, 2 gallons olive oil,* # 2 canning salt,
*4 jars Miracle Whip, *8 # 4 peanut butter, *3 jars tahini, *6 pkgs.
Jell-O, *5 pkgs liquid pectin,* 5 boxes powdered pectin, *10 boxes
lemon herb tea, * 12 large cans coffee,* 3 large chili powder, * 3
large paprika, * 2 large black pepper, * #.50 ground red pepper, * 1
bottle hot sauce, * 2 large ground cumin, * 2 large minced onions,
*#1 ground ginger, * 3 - 1 oz. bottles of almond extract,* 1 bottle
of Marsala wine, *1 bottle of good whiskey

Flour Grains & Beans
*#50 lbs. all-purpose flour, *#20 bread flour, *1 box cake flour, *5
large boxes oatmeal, * 1 box Cream of Wheat, *# 3 barley pearls,
* #10 black beans, * #6 red kidney beans, * #5 chick peas, * #3
lentils, #50 wheat berries, * # 1 sesame seeds

Baking Supplies & Sugars
*# 85 lbs. cane sugar, 3 quarts maple syrup, * 3 cans cooking
spray, * 2 large bags of instant yeast, * #6 confectioners' sugar, * 1
can vegetable shortening, *# 8 brown sugar, * # 3 shredded
coconut, * # 3 chopped dates, * #2 dried apricots, * # 2
butterscotch chips

Household Supplies
6 boxes large kitchen matches, 10 gallons bleach,* 5 gallons white
vinegar, * 1 box borax,* 1 gallon ammonia, * 30 vacuum bags, * 8 large
boxes Tide, *6 large bottles Downy fabric softener, * 1 box dryer sheets,*
14 large bottles Dawn dish soap, * 9 large boxes dishwasher detergent, * 3
bottles toilet cleaner, * 12 cans Bar Keeper Friend, * 5 bottles tub & tile
cleaner, * 50 rolls toilet paper, * 24 rolls paper towels,* 4 boxes Kleenex,*
6 large pkgs. paper napkins,* 80 tall white kitchen garbage bags,* 1 box -
30 gallon size garbage bags, * 1 box wax paper,* 2 boxes aluminum foil,
*mouse traps, * rat poison

Medicine & Personal Care
*4 large bags disposable razors, *4 cans shaving cream, * 4 large
boxes Q- Tips, * 8 tubes toothpaste, * 6 boxes dental floss,* 8
large Listerine, *4 roll-on-deodorant, * 4 mascara, * 1 brown eye
pencil, * 3 jars old lady eye cream,* 2 large Tylenol, * 2 large
aspirin, * 2 tubes hydrocortisone cream, *2 large Vaseline, * 3
large cold cream, *5 hand cream, * 3 cans hair spray, * 3 large
shampoo, * 4 bags cough drops (maybe make horehound drops
too)* 2 bottles cough expectorant, *2 bottles cough suppressant,* 2

The Complete Granny Miller

bottles Pepto Bismol, bar soap, bath oil, * essential oils for soap &
bath

How to Plan For a Year's Worth of Food and Supplies

My tool for planning a year's worth of food and household supplies is a
detailed and complete household inventory. For over 25 years, I've done a
yearly household inventory and recommend the practice to anyone seriously
interested in a more self-reliant life. By now, I'm pretty proficient at
estimating the quantities of food and household supplies for my family for a
year or longer, due to sheer applied experience. But when I was younger,
there was no such thing as the Internet. Back then, basic household
management information was obtained through home economics books
borrowed from the local public library, recommendations, oral tradition,
and histories from the two preceding generations of women.

In the beginning, when I started planning for a year's worth of household
needs, I had no idea how many rolls of toilet paper, green beans, Q-Tips,
pounds of hamburger, or cans of coffee my home used in a year. I figured
this out by keeping a diary on a wall calendar with a pen. Every time I
opened a package of chicken, a box of laundry detergent, a bottle of aspirin,
or a jar of peanut butter, I wrote it down on the calendar. I was committed
and consistent in recording all food, supplies, and everyday household
items. After the first year, I had a good understanding of what I was and
wasn't using, and how much of each item I consumed. It took me about two
or three years of paying close attention and tracking what was actually
happening in my home. This information helped me make informed
guesses, letting me plan confidently for a year's worth of food and supplies
without wasting money. I needed to learn about and understand my family's
unique consumption patterns.

I'm sure that there may be other ways or better ways to glean that type of
household information. But the simple calendar diary is what worked for me
and got me started on the road to sensible planning for a year's worth of
food.

Take Inventory Every Year

With a yearly inventory, I'm able to notice possible patterns of changing food
ways, behaviors, or brand preferences, and I can make the necessary
adjustments. The information I gain provides the foundation for a master
household and pantry shopping list. By planning and purchasing household
supplies and specific foodstuffs ahead of time, I save time, money, and gas
since I don't have to make weekly trips to the grocery store. Most of the
time, the only shopping I ever do between November and June is in my
cellar with a laundry basket.

Once I have a list to work with, I know precisely what fruits and vegetables
I need to grow for any particular year, and how much to home can or freeze.

The Complete Granny Miller

My list also informs me of the type or quantity of livestock or poultry needed to be raised or slaughtered in any given year. I use my master list to help me purchase and store food and household supplies that I can't or don't want to produce myself. Items such as light bulbs, paper towels, dishwasher soap, lipstick, olive oil, and brown sugar are staples that I regularly purchase. By spreading the purchase of store-bought items out over five months, my monthly budget is less impacted, and the acquisition and storage are easily accomplished.

The Complete Granny Miller

Chapter 6

"The ultimate goal of farming is not the growing of crops, but the cultivation and perfection of human beings." - Masanobu Fukuoka.

Homestead Animals

For many people, the addition of animals to a homestead or a small holding is what transforms a simple backyard vegetable garden into a garden farm. All livestock, whether they are large or small, define a garden farm and make a household more viable, self-reliant, and self-supporting.

The following section is not a comprehensive how-to for poultry or livestock owners.
There are plenty of other good books for that. Instead, this chapter is a collection of past posts from GRANNY MILLER. It presents the reader with snippets of basic livestock information and advice that can sometimes be hard to come by. It's heavy with sheep and cattle information, and there's limited information about everything else. Seems I had a preference for cattle and sheep back in my blogging days.

Some Good Advice

Large animal veterinarians are scarce in many areas of the country. Often, when a large animal veterinarian does make a farm call, the service can be expensive, and there's never a guarantee that your animal will survive with or without veterinary treatment. Small holders and homesteaders are obliged to do the best they can for the animals in their care. Many years ago, an old dairy herdsman told me,

"When you got animals, you got troubles."

Well, that really isn't exactly true. But when you have problems in the barn, sometimes it sure seems that way. Living things by their nature are unpredictable. The truth is, life is precarious and capricious. Farmers know this probably better than anyone. So, a few words of wisdom about livestock:
- Livestock are not pets. This doesn't mean that you shouldn't love your animals or display kindness and human affection. But what it does mean is that every animal life has a price.

- Livestock that are meant for food are expendable. Never go into debt for an individual animal.
- Veterinarians are human just like you and me. Veterinary medicine has limitations. Just because you call the vet doesn't mean things will work out.
- Never call a veterinarian to attend your animals without a price limit for treatment. Believe me, veterinarians know better than you or I that every animal life has a price. Be aware that you will often end up with a dead animal and a hefty bill. It's not the veterinarian's fault. It's just the way things are.
- Never bring home or spend money on an animal that you aren't prepared to see dead in the barn the very next morning. Livestock sometimes have a way of up and dying without giving a warning.

In fact, there's an old saying about that. It's in regard to goats and sheep.

"Sheep are born looking for an excuse to die. Goats take the first chance they get."

Farmers and professional gamblers have a lot more in common than you might first suppose.

CHICKENS

Feather Loss in Chickens

About once a year, chickens go through a molt. Molting is the time when a chicken sheds old feathers and grows new ones. It is an entirely natural occurrence and takes anywhere from 5 to 12 weeks to complete. In hens, molting usually follows a period of heavy egg production. Hens will not lay eggs at all or will lay eggs sporadically during their molt. Laying hens usually fall into two groups - late molters and early molters.

Hens known as late molters will lay eggs on average for 12 to 14 months before they begin to molt. Late molters are generally the better laying hens, and they will often have a more ragged and tattered appearance during their molt. Hens known as early molters sometimes begin to molt after only a few months of egg production. Early molters take longer to complete their molt and are often poor layers. They can have a fuller feathered appearance and don't look as moth-eaten. Early molters often will shed only a few feathers at a time.

Autumn is the traditional time of year when molting occurs due to a decrease in daylight hours. During the molt period, feathers are lost in a predictable sequence. Feathers are lost from the head first; then followed by those on the neck, the breast, the body, the wings, and lastly the tail.

Treading Marks

Sometimes the loss of feathers on the back of a hen is from more than just molting. It's called a treading mark. A treading mark develops when a hen becomes a particular favorite with a rooster or a group of roosters. During copulation or mating, called treading in chickens, the rooster's feet sometimes will tear feathers from a hen's back as he moves his feet quickly across her back while he is on top of her. Some hens are trodden so often that they will develop a big bare spot on their back. A hen will usually grow her feathers again if she can manage to stay away from the rooster.

Baby Chick Care

I start my chicks late in the year to take advantage of the warmer weather. They are usually shipped in a cardboard box and are typically quite thirsty upon arrival. I take them out of the box individually and help them to take their first drink of water. For their first drink, I mix a commercial vitamin solution into the water to keep them healthy. A vitamin solution is especially important when raising meat chickens or any of the heavy breeds, as they grow so rapidly.

Baby chicks need to be kept warm. For most homesteaders, that is usually achieved with a heat lamp or light bulb. For those who are non-electric, kerosene brooders are available on the market. But kerosene brooders are expensive and can be fussy to operate. It is easier to keep chicks warm when the daytime temperatures are already above 80°F and seldom fall past the mid-50s at night. Here's the advice I always give to chicken beginners. The most important thing is to ensure your baby chicks stay warm and dry. A heat lamp hung above a draft-free, enclosed brooding area works well for this. For the first week of life, 95°F to 90°F is just about right. Reduce the temperature by 5°F each week until they are fully feathered. A thermometer is a helpful tool for monitoring temperature. However, you can also tell if they are warm enough by their behavior. If the chicks are huddled tightly under the heat lamp, they are too cold. Lower the lamp for more heat. If they are far from the lamp, they are trying to escape the heat; they are too hot. When the chicks are comfortable and resting, they tend to spread out around the lamp, sometimes with their wings stretched out or on their side. Keep young chicks clean, fed, watered, and not too crowded, and they'll be okay. I always give my young chicks a few handfuls of freshly pulled-up grass with dirt clinging to the roots a couple of times a day. I think it keeps them healthy and gives them something to do until they are big enough to go into a chicken tractor or free range. And just so you know, chick chirping is perfectly normal. It sometimes means they are hungry. But if the food and water are in front of them, it just means that they're happy.

The Complete Granny Miller

Pick the Best Day for Hatching Eggs

I firmly believe in agricultural traditions and folk wisdom because much of what I learned about homesteading was passed down from two generations of garden farmers before me. Following their advice has led to greater success and fewer failures in homesteading. One valuable piece of advice I received from more experienced garden farmers was about the best time for setting or incubating chicken eggs.

The most favorable time for setting eggs under a broody hen or in an incubator is 21 days before a waxing moon in the zodiac moon sign of Cancer. In order to determine what day that would be, you'll need an almanac for the current year. All good almanacs feature tables or charts that map the moon's course through the zodiac.

Chicken eggs need 21 days to hatch. So, a quick look in any current almanac will find days that the moon will be in the sign of Cancer and will also be waxing. Most years, there will be a couple of days when this occurs during the light (waxing) phase of the moon. All that is necessary is to pick a lunar Cancer day and then count backwards 21 days. Whatever day that happens to be is the day to begin to incubate the clutch of eggs. That day counts as Day 1.

If, for some reason, a waxing Cancer day is inconvenient, a day that a waxing moon falls in the signs of Scorpio or Pisces would be a second-best choice. Chicks that are hatched during a waxing Cancer moon tend to hatch with fewer problems and grow faster.

SHEEP & GOATS

After chickens or meat rabbits, goats or sheep are typically the next livestock that homesteaders and garden farmers will acquire. Both goats and sheep are classified as small ruminants. They have needs that are somewhat different from those of chickens, rabbits, or other small livestock. When dairy goats or sheep are added to the garden farm, food variety is increased exponentially. Both goats and sheep provide the foundation, understanding, and experience necessary for larger livestock, such as horses, beef, or dairy cattle.

The Homestead Dairy

In the past few years, there's been a growing movement for new homesteaders or inexperienced small holders to begin with a family cow for a dairy animal instead of a good dairy goat. On one hand, I don't blame people for not drinking goat milk or eating chevon or capretto (goat meat), because I don't eat or drink goat products either. But on the other hand, I'm troubled by so many ill-equipped, inexperienced, and ignorant people buying dairy cows who have no business owning them. It's my opinion that

The Complete Granny Miller

a good dairy goat, and not a cow, should always be the first dairy animal for a homestead.

The Right Dairy Animal

My husband and I have a mixed marriage. He's a goat person, and I am not. If it were up to him, this farm would be positively overrun with dairy goats. He loves them!

Goats are probably the smartest and most personable of all livestock. They are naturally charming and entertaining - especially the kids. A smart goat is about as intelligent as a really dumb dog. Goats will find any and every way to avoid confinement, and that makes for trouble. You'll need fences that will hold water to keep a goat! To my way of thinking, goats are more trouble than they're worth.

While my husband and I don't agree about goats, we do agree that milk is a foundational food for a homestead. We also agree that goats are probably best for people of modest means. Most families don't need a cow that will produce 3 to 8 gallons of milk a day. A good dairy goat is perfect for small families or couples. But where my husband and I part ways is regarding the taste of goat milk. My husband can't tell the difference between goat milk and cow milk. But I most certainly can. I never drink goat milk because of the taste.

Goat people almost always have a negative reaction and take exception to that statement. Dairy goat aficionados will insist that goat milk is fragile and must be handled with care to ensure that there's no off taste to the milk. And that's true. Goat milk will readily pick up an off flavor if it's not handled correctly. In fact, when it comes to handling fresh goat milk, I've found that the best way to cool down a small amount of milk fast is with a Cuisinart ice cream maker. The freezer bowl cools the milk faster than a standard bulk tank, resulting in perfect Grade A milk every time.

But I think there is something more to the goat milk story than simply milk management. I believe there's some type of genetic component in the ability to taste 'goatyness' in fluid milk. It's kind of like the ability to roll your tongue. Not everybody can roll their tongue. Maybe the goat milk taste thing is genetic. Some people may have the genes to taste the goat in the milk, and others can't. Many people can taste the difference between Grade A goat milk and cow's milk.

In general, if you are like me and hate the taste of goat milk, you may be better off with an extremely low-producing dairy cow. I own dairy cows, but don't keep a family cow here on the farm. I no longer have children at home, and I don't make cheese regularly. But if I had children at home and was up for making cheese every couple of days or so, I'd indeed keep a cow. But before I would ever consider a family cow, I'd want a separate area to process the milk. Dealing with three to eight gallons of milk every day for up to 300 days a year presents a huge storage problem in most kitchens. Not to mention that serious cheese-making can make a big, greasy mess. Handling gallons of milk day in and day out is a tremendous amount

136

of work and is time-consuming. In the past, farmwives found the time to make butter and cheese because they had plenty of help, whether from children or a hired girl.

A cow, unlike a goat, represents a substantial financial investment and a significant personal commitment. You'd be hard pressed in my part of Pennsylvania to buy a cow for less than $700 (2010 price). Once you have that cow, she needs to be housed, fed, cleaned up after, and receive periodic veterinary care. Cows, like all mammals, give milk only after giving birth to offspring. A cow will need to be bred in order to give milk. Most single cow families will elect to borrow a bull, ship the cow to a neighbor who has a bull, or have the cow artificially inseminated. All three of those options have benefits and drawbacks.

My advice to most people considering a home dairy animal is this: Try fresh goat milk to see if you like it. If you do - great! Get the nicest goat you can afford and enjoy her and her milk. Good-quality dairy goats range in price from upwards of $150, with about $175 to $200 being the current going rate (as of 2010). A dairy goat is the very best choice for a small homestead or backyard garden farm. She can be easily bred by borrowing a buck goat or by taking her to a neighbor with a buck goat when she is in heat. A goat is much easier to transport than a cow. A high-quality dairy goat will often produce well over a gallon of milk a day. All babies, human or animal, thrive on goat milk. For years, I tolerated goats solely to provide milk for bottle lambs and feeder calves. It's cheaper than commercial milk replacer. And nothing I know of will fatten a pig faster than corn and goat milk. You'll have some trouble making butter with goat milk because the cream won't readily separate. But with a goat, you will have the means and the milk to make the best fudge and feta cheese you've ever eaten. Not to mention luxurious skin-soothing goat milk soap.

But if you're like me and can't endure the taste of goat milk, you only have a couple of realistic options. If you have a large family, good pasture, and the time and inclination to make cheese and butter, a cow is the way to go. Try to buy a low-producing dairy cow that's in calf or already lactating. If she's in calf, once she freshens, let her keep her calf for half the day; usually during the night. That way, you don't have to milk twice a day, and you still get lots of milk. If you get a bull calf, you'll have a calf that will grow for meat. If it's a heifer calf, you can sell her once she's weaned.

If money is no object and you already have some knowledge of cattle, Dexter and Kerry cows are both known for being extremely low milk producers, making them potentially good options if you can find them. However, be aware that a Dexter cow often does not produce enough milk for both the kitchen and her calf. It's pretty much either milk for your family or milk for her calf - but not both. You may have to bottle-feed the calf (preferably with goat milk) or wean the calf early. You can give a

The Complete Granny Miller

supplemental bottle to the calf after it is about 6 weeks old, or simply go without milk until the calf is weaned.

Often, a small commercial dairy cow that is being culled from a dairy herd can be found for a reasonable price. The right cull dairy cow is what I recommend to families on a budget looking for a family cow. Many perfect dairy cows are culled because they are considered to be low commercial milk producers. But these cows will work well as a family milk cow. Sometimes dairy cows are culled from a herd because they have one or two bad quarters or are susceptible to chronic mastitis. Both of those conditions can often be made manageable on a small homestead with careful attention. For small families, older individuals, or those who don't need to milk 300 days a year, I believe a cow share is the best option for high-quality milk. If a cow share is not possible, perhaps you can find a neighbor with a cow that will sell you milk.

Sheep

I bought my first sheep well over 25 years ago. Sheep were a natural outgrowth of my interest in hand spinning and weaving. My first flock of sheep was purchased with the profits from a small, seasonal commercial decorating business that I started. Here's what happened.

The year I bought my sheep, I risked $50 for pumpkin, gourd, and Indian corn seed. I didn't have a lot of money back then, but I had saved some money from the part-time school bus driving I'd been doing. I planted the seeds in the spring, and that year turned out to be an outstanding year for corn and pumpkins. By mid-September, I had more than enough pumpkins and Indian corn for a small business enterprise. I printed up a price list, and that fall I made cold sales calls to every local business within a 25-mile radius. The service I offered was straightforward. I'd supply pumpkins, corn stalks, and Indian corn, and then beautifully decorate a given business for the fall season.

That September and October, I decorated banks, restaurants, gift shops, convenience stores, and gas stations. My clients were so pleased with the service I provided that I ended up with repeat clients for Christmas. My seasonal decorating business continued for another year or so until I finally grew disenchanted. In fact, I completely burned out on decorating anything. It's one of the reasons, still to this day, I'm not wild about electric Christmas lights or artificial Christmas trees. But that's another story, and I digress. The point of my tale is that capitalism, hard work, and having a good idea before its time can pay off.

Sheep, in many ways, are similar to cattle. The lessons that I learned by keeping sheep were easily transferred to cattle a few years later when I began a small beef herd.

The Complete Granny Miller

Advice About Sheep

The advice I always give to folks who are thinking about starting with sheep is the same advice that was given to me well over 25 years ago.

"Don't buy into sheep; grow into sheep."

That's still pretty good advice. If you have never raised sheep before, it's best to take your time and go slow. Sheep will take a while to learn about and understand if you have never been around them before. Observation and daily interaction with any new species of animal is always the best teacher. Books can only impart so much information.

What Kind of Sheep?
Well, that depends. What do you want to do with your sheep? Are you a handspinner or avid knitter who wants homegrown wool? That's how I started my first flock. Or perhaps you see an ethnic market opportunity and would like to direct-sell holiday lambs? If that's the case, you'll want a different breed of sheep.

First and foremost, never buy just one sheep that will be kept as an 'only'. Except for dogs and cats, most domestic animal species need another of their kind in order to thrive. With sheep, it is vital. That's because sheep are prey animals. Prey animals are almost always flock or herd animals and feel safest and most content within a group. If you simply must have that adorable orphaned bottle lamb, then make sure to have another animal to keep it company: a goat, cow, horse, or any other animal with cloven hooves. Just don't keep a sheep alone.

Choose the breed that is right for you and your place. All farms are unique habitats. Each farm or homestead has different needs and goals. A breed of sheep that may work well and thrive on my farm may not necessarily do well for my neighbor. Pay attention to what types of sheep are being raised in your area. There are usually good reasons that a specific breed is popular or has been chosen in a particular location. Different breeds have different strengths and weaknesses.

For example, where I live, the ground tends to be wet. Merino or Dorset sheep don't do as well in this area as some other breeds do. Fine wool breeds seem to be more susceptible to foot scald or foot rot than other breeds. Foot problems are aggravated by wet or soggy ground. Cheviot sheep are a better choice for my farm. Happily, there are always exceptions, and good management can overcome certain natural tendencies. If you have your heart set on a particular breed, by all means, get that breed. Sheep are adaptable creatures. You just need to be aware that there are differences between breeds.

Where to Buy Sheep
Stay away from the livestock sale barn when buying sheep. Sometimes

The Complete Granny Miller

there are great bargains to be had there. But for the novice, more often than not, you are buying someone else's problem. And while I'm on the subject of the sale or auction barn, please remember this: change your boots and shoes, or disinfect them with a mild Clorox solution, when you return to your farm from the sale barn. Or any place where farm animals have been kept. Do not go into your barn or pasture without doing this. You'll save yourself a load of foot rot and other trouble by this simple precaution.

Keep in mind the adage:

"You get what you pay for."

A young, yearling ewe with a fancy pedigree will cost considerably more than a crossbred 5-year-old ewe. There are often good deals to be found in the fall or winter by buying the cull ewes from a reputable breeder. If I were a beginner starting from scratch, this is how I would approach it. You will not have as much money tied up in your stock, and your financial loses will be less should you change your mind about keeping sheep. A bonus with an older, proven ewe is that she's not the worry that a young ewe is. A 5- or 6-year-old ewe will still have a few good years left in her. An older ewe will not be as nervous as a first-time mother or as a maiden ewe may be. If you are a beginner with sheep, you may lack confidence and be nervous enough without any help. An older, experienced ewe will teach you more about sheep and lambing than anyone could hope to. Trust her. She knows what she's doing.

Try to find a reputable breeder. Look for one that has been involved with sheep production for more than a few years. Rely on your instincts when you first meet a breeder. Livestock breeders, for the most part, are honest people. They love their animals and want what is best for them. Dedicated breeders invest a tremendous amount of time, effort, and money in trying to improve their chosen breed. However, do keep in mind that fancy ribbons and show titles, with a high price and a snooty attitude, don't necessarily mean you have found a good breeder. It can often mean you have found someone with a big ego, who's working the show circuit in hopes to make back a poor investment, or to recoup poor economic choices.

Whenever I go to look at sheep, I pay close attention to the condition of the fences and barn. You should do the same. The facilities don't have to be slick and new. Old and make-do is just fine and is often the most sensible and typical arrangement. If things are in good repair and well-tended, you can be pretty confident that the animals are being well cared for, too. Of course, there are exceptions to this rule. Many people take good care of their animals, but they often fall behind on repairs. But you can be sure that if the fences are down, manure in the barn is piled up almost to the roof, and there are buckets and trash thrown everywhere, things are probably not what they should be.

When you first go into a barn or to wherever sheep are being kept, notice what you see and pay attention to the air. The air should smell clean and not heavy with ammonia. There should be plenty of fresh, dry bedding. Be on

The Complete Granny Miller

the lookout for sick sheep: ears down - standing off away from the flock; pools of diarrhea, sheep with dirty, wet rear ends; or piles and piles of manure. Pay attention to how the sheep are being watered and whether or not the water is fresh and clean. If hay is being fed, note whether it is being fed from feeders or off the ground. Pay attention to the number of barn cats. Every barn needs some cats. But if the barn is overrun with cats, toxoplasmosis could be an issue.

Familiarize yourself with the breed characteristics of your chosen breed. When looking over the sheep being offered for sale, keep that in mind. Don't hesitate to ask the breeder for input or to help you choose the best of the lot. Most reputable breeders are more than happy to help with this. Try to choose ewes with a flat, straight topline, sturdy legs and strong feet, a sound udder, and a good mouth. In a proven ewe, ask about her lambing history. Try and choose a ewe that twins regularly and has never rejected her lamb. Good mothering is inherited in sheep. This trait will be passed along for generations. Avoid ewes that have been bottle lambs or orphan lambs if at all possible. They do not typically make the best mothers.

When Buying a Ram

There's an old saying:

"A ram is half the flock."

A ram's contribution to a flock will be apparent for many years. Expect to pay two to three times as much for a good 2-year-old ram as for a ewe. Make sure to get in writing as a condition of the sale that the breeder will accept the ram back if he proves unfit to breed or is infertile. Most reputable breeders will mention this without you having to bring it up. In choosing a ram, examine teeth, wool, and all four feet and limbs carefully. Reject any ram with excessively sloping pasterns or splay feet. Rams should have a good scrotal circumference, and the scrotum should appear clean, covered with soft skin, and feel heavy when it is lifted. In general, the larger the testicles, the more sperm and male hormones are produced. This is important for a ram because he usually needs to be able to breed many ewes within a short amount of time.

Bringing New Sheep Home

When you bring your new sheep home, keep them confined for about a week or ten days so that they can get accustomed to you and the new surroundings. Try and buy three days' worth of feed and hay from the seller or breeder. It's imperative that your new sheep not have a radical change in diet. They have already been upset and stressed by the move. You don't want them to get sick.

Feed them lightly at their first feeding once you get them home. On the first day, feed them about 75% of the daily ration of hay and feed that they are used to. The next day, mix about 25% of the new feed or hay (from your place) with their old feed. Over the next two or three days, gradually

The Complete Granny Miller

transition them to your feed and hay. After five days, if you wish, turn the sheep out into new pasture. Feed them hay first and then turn them out for about an hour before the sun goes down. Repeat this the following evening and increase the time to about two hours. On the third day, give them plenty of hay and turn them out for half a day; on the fourth day, turn them out for the entire day. It takes a while for the bacteria in a sheep's rumen to become adjusted to changes in the diet. Remember: A healthy rumen is a healthy sheep.

Teach Them to Come When Called
Last, but not least, it's valuable to teach all farm animals to come when called. This is easily accomplished by calling your special call (make one up) every time you feed them. After ten days of this training, they will come whenever called. If they ever get out of the fence or you have to load them onto a trailer, you'll be thankful that you took the trouble to teach them to come when called.

Healthy Sheep Feet
Over the years, I've noticed that sheep with black hooves tend to do better on wet ground than sheep with lighter colored hooves. Wet ground can lead to foot scald in sheep, leaving the hoof susceptible to foot rot. Some sheep breeds with light-colored hooves include Dorset, Merino, Finns, and Polypays. Sheep with light colored hooves do well in dry climates. So, the next time you're thinking about buying sheep, keep in mind the kind of pasture or ground the sheep will be standing on. You'll save yourself and your sheep a lot of trouble.

Foot Scald & Foot Rot
All foot problems in sheep should be investigated and treated as soon as possible. There are lots of different reasons for sore feet, and foot rot isn't always necessarily one of them. But lameness in more than one sheep is a red flag. Two or more limping sheep is cause for an immediate evaluation of the entire flock. Don't ignore the warning sign or waste any time investigating the cause.

Foot rot is an infectious and contagious disease in sheep. It is caused by the interaction of two different anaerobic bacteria: *Dichelobacter nodosus* and *Fusobacterium necrophorum*. Foot scald, while not as serious as foot rot, can be just as painful. If not treated promptly, it can develop into foot rot. Foot scald occurs from the opportunistic interaction of Fusobacterium necrophorum and Trueperella pyogenes during wet weather or damp conditions. For clarity, we'll refer to these bacteria as B1, B2, and B3. We will also make the words 'hoof' and 'foot,' 'toe' and 'cleat,' interchangeable. F. necrophorum (B1) is a normal bacterium found in the digestive tracts of sheep and other ruminants, and it can survive up to ten months in soil. In wet weather, F. necrophorum (B1) can interact with T. pyogenes (B3) to cause foot scald. Foot scald is an infection of the skin between the toes of the hoof and serves as a precursor to foot rot. Mud acts as an ideal vector for these bacteria, easily packing into and squishing between the toes when sheep are on wet ground. Keeping sheep on soggy ground promotes foot scald, even in sheep not genetically predisposed. More about genetic susceptibility and foot rot will be discussed later. Once the hoof is compromised by foot scald, D. nodosus (B2) can invade the foot. D. nodosus (B2) produces an enzyme that destroys the connective tissue between the horn (the hard part) and the tissue of the foot, allowing bacteria to migrate under the horn of the hoof. Once beneath the hoof, the bacteria reproduce rapidly in the anaerobic environment, protected from air by the hoof's tough covering. This difficulty in reaching the bacteria is one reason foot rot is so hard to treat.

In my experience, the most effective treatment for foot rot or foot scald is a combination of isolation, foot baths, hoof trimming, and the use of Liquamycin (LA 200). LA -200 is an injectable oxytetracycline solution. Here's what works for me:

First, isolate the sheep into two groups: the limpers and the non-limpers. The two groups will be treated differently.

The Non-Limpers

Set up a foot bath for the non-limpers and make them walk through it. Hold the sheep in the foot bath for a couple of minutes if possible. Use a gate or hurdle if need be. After the foot bath, move the sheep immediately onto fresh, clean, dry ground. Watch the sheep carefully over the next week to ensure that none of them are limping, even slightly. If you find another limper, move it immediately to the other group.

There are all kinds of products for foot baths. But a mild Clorox solution is what I use and recommend. Chlorine bleach is cheap and is readily available. The foot bath mix is a 10:1 ratio of water to Clorox. If you don't have a special trough for a foot bath or only have a few sheep, a plastic bucket with the bleach solution will work fine as an individual or single hoof foot bath. Just stand each hoof in the Clorox solution for about 1 minute. Don't forget to wash your footwear too. Boots and shoes that have walked over contaminated ground will spread foot rot. That's why it is so important to change your shoes before you go into the barn or pasture after you have returned from the sale barn, auction, or from visiting another farm.

The Limpers

Confine all the limpers on a heavy bedding of fresh, clean straw. Proceed with the treatment below, depending on whether or not you have foot scald or foot rot.

For Foot Scald

Examine and trim all four hooves. Not just the lame one. A hoof with foot scald usually looks moist between the cleats (toes) and is slightly reddened.

The Complete Granny Miller

Clean out any packed-in debris and wash between the toes. Next, apply a light coating of LA 200 between the toes. Do this with a syringe that has had the needle removed. To fill the syringe, draw the LA 200 up into the syringe with the needle attached. Then remove the needle. LA 200 is runny, making it hard to control where it goes. Drip and drizzle the LA 200 over the entire interior wall and skin between the toes. Any area that appears moist or inflamed should be well covered and coated. After treatment, keep the sheep on clean, dry straw for a week. A difference in limping will probably be noticed within 24 to 36 hours. LA 200 applied directly to the scald is a highly effective treatment for foot scald.

For Foot Rot
A hoof with foot rot will have a foul smell. You can't miss it. The hoof may appear crumbly and soft, or it may peel away. At times, the bulb at the back of the heel will be oversoft, white, or spongy. Sometimes the hoof will be bloody.
Clean and trim all four hooves. Try and trim the hoof so that a line of white or faint light pink can be seen around the fresh trimmed rim of the hoof. Try not to over-trim the hoof, or it will bleed. Using the same treatment for foot scald, apply LA 200 directly to the foot with a syringe that has had the needle removed. Next, with a clean sterile needle attached to the syringe, administer a 10 cc IM injection of LA 200 divided into two 5 cc doses, into two different injection site locations. Be aware that LA 200 is a painful injection. It burns. Expect bucking, rearing up, and general carrying on from your sheep when you administer it.
Keep the foot-rot-treated sheep on dry, clean bedding for at least 7 to 10 days before they are turned out again. Do not let the sheep return to infected pastures or paddocks. *D. nodosus* (B2) can live on pasture for 14 to 22 days. Once the disease mechanisms involved are understood, both foot rot and foot scald can be prevented and eradicated with careful flock management.

Genetic Susceptibility
Any breed of sheep can end up lame given the right conditions, mismanagement, and bacteria. As a rule, sheep with black hooves are less prone to foot rot. Fine wool breeds or sheep with white, pink, or striped hooves are much more prone to foot problems and don't do as well on wet ground. Polypays, Merino, Dorset, Finn, Rambouillet, and Columbia breeds are sheep that have given me foot problems in the past. I think if my farm were drier, they would do better for me. Some individual sheep seem to have a genetic susceptibility to foot rot. I have often wondered if it's not some type of innate immune system problem. I just don't know. But what I do know is that sheep that show a susceptibility to foot rot or foot scald, and don't respond well to treatment, should be culled from the flock and their offspring not kept for breeding. Don't breed problem sheep.

The Complete Granny Miller

Ram Marking Harness

One of the most practical sheep management tools during breeding season is the use of a ram marking harness. A ram marking harness is a small harness typically made of nylon or leather that holds a square-colored crayon. The way that a marking harness works is that a colored crayon is attached to the front part of the harness and is centered over the ram's brisket or chest area. When a ram with a properly fitted harness and a temperature-correct crayon mounts a ewe, the color from the crayon is transferred onto her rump. By monitoring the backsides of a group of sheep, it's easy to determine which ewes have been bred. Most ewes will deliver lambs approximately 145 to 149 days after being serviced by a ram. A note on a calendar indicating the service day records the breeding and helps to calculate when a given ewe can be expected to lamb.

Marking harnesses are available in two or three sizes to accommodate the various breeds of sheep. Marking crayons are manufactured in differing degrees of hardness. Marking crayons are temperature-sensitive, and the proper crayon should be selected according to the expected weather conditions during the breeding season. Hard crayons are used during hot weather for temperatures 85°F and above. Medium crayons are for mild weather, 60°F to 85°F. Soft crayons are for use during cool-weather breeding when temperatures are less than 60°F. Marking crayons come in several colors and are used in a few different ways.

One way to use a marking harness and crayons is when two or more rams are working in a single group of ewes. When two rams are working together, each ram wears a different color crayon marker on its harness. This system is typically used for 50 or more ewes.

Another way that crayon marking is helpful is when space is limited. By observing the backsides of ewes 4 or 5 times a day, a ewe can be removed from the group after she has been marked if necessary. When ewes are run with two or more rams, there's no way to determine the sire of her lamb(s) if two or more colors have marked a ewe. With grade sheep, it's not a problem, and the extra ram power is an advantage. But this method should never be used for registered sheep or when the offspring may need to be papered.

When using a marking harness, it's essential to be able to distinguish betweena jump smear and a good, solid breeding mark. Often, a ram will attempt to mount a ewe that will not stand for him, and he will leave a faint smear or streak of color on her backside or flank. When a ewe has been honestly bred by a ram wearing a marking harness and with the correct crayon, the mark on a ewe's backside is distinct.

Different color crayons are helpful when trying to determine if ewes are pregnant. Or if ewes need to be re-bred, or to determine if a ram is possibly sterile. Most sheep are naturally polyestrous short-day breeders. Active estrus in ewes lasts approximately 24-36 hours. It is the only time that a ewe will stand to be mounted by a ram. Without human intervention like sponging (sponging is a method of altering a ewe's natural cycle by using hormones saturated on a tampon-like sponge that is inserted vaginally for a

The Complete Granny Miller

few days and then removed), a ewe will begin a normal estrus cycle every 15 - 17 days when daylight hours begin to decrease in the autumn.

By changing the crayon color on the harness every 16 to 18 days, it is possible to observe if a particular ram is re-breeding ewes. If he is re-breeding most of the ewes, he probably was temporarily sterile when he was first turned in with them. Some rams are especially heat-sensitive and can become sterile or have a low sperm cell count for several weeks after a summer heat wave. Most heat-sensitive rams will recover fully once the weather turns cool again. A ram can also be made temporarily less potent if he is asked to service too many ewes at one time. Thirty to thirty-five ewes are about as many as a fully mature and experienced ram can handle. There are different reasons why a ewe may not get with a lamb and settle after the first tupping (tupping is the term for copulation in sheep). It happens often, some years more so than others. It's not unusual in a large group for a ram to have to re-service a few perfectly normal ewes during subsequent estrus cycles.

Back in the old days, before there were marking harnesses and colored crayons, sheep breeders would paint the brisket of their rams daily with a mixture of linseed oil and different colored pigments or lamp black (soot). It was a low-tech and much messier way to record sheep breeding activity. But it sure enough got the job done.

Lambing & Kidding Time

Most of the time, sheep and goats require no human intervention during lambing and kidding. But occasionally, a problem will pop up after delivery. The next part of this chapter addresses two common issues that can arise after lambing or kidding.

Retained Placenta

Typically, after lambing or kidding, a ewe or nanny will expel the afterbirth or placenta within an hour or two. But sometimes the placenta can be stubborn about being released during the cleansing phase of lambing or kidding. In sheep and goats, a placenta that does not evacuate the uterus after 12 hours or so after delivery is known as a retained placenta.

There are several possible reasons for retained placenta in small ruminants.

Too much grain, low-quality hay, an over-large lamb or kid, lack of exercise, nutritional deficiencies, premature birth, stillbirth, abortion, and infection are all associated with retained placenta in livestock. A retained placenta is usually no cause for alarm as long as a few simple guidelines are followed.

It's safest not to try and manually remove the placenta. Often, the placenta is completely retained, and there is no sign of it. But sometimes the retained placenta will be seen hanging out of the vulva. After 12 hours or so, little harm will be done by gently testing the adhesiveness of the placenta. In such cases, if the placenta doesn't readily flop out of the ewe or nanny after a slight tug, leave it alone. It's important to remember that the placenta is attached to the uterine wall by disk-shaped cotyledons. If you try to pull the placenta away from the uterus before it is ready to be shed, you can injure the ewe or doe and run the risk of adversely affecting her future pregnancies. Pulling on the placenta also increases the chances of bleeding and infection.

The best course of action for a retained placenta is to prophylactically protect the ewe or doe with an antibiotic and keep a watchful eye on her.

In ewes or does that have retained their placenta, I use 10 cc to 20 cc of injectable penicillin (Penicillin G Procaine – 300,000 units per ml) via a SQ or IM injection every 48 hours until the placenta is released. Most often, the placenta is sloughed away within 3 to 5 days, and the ewe or doe will go on with life as if nothing happened.

But if the doe or ewe should go off her feed, she may need an injection of dexamethasone as a supportive therapy. Dexamethasone is a synthetic analogue of prednisolone. It has a similar but more potent anti-inflammatory therapeutic action than prednisolone and is available only on the order of a licensed veterinarian.

Hypothermic Lambs and Kid Goats

Lambs and kid goats can withstand quite a bit of cold, provided they are well started, stay dry, and receive plenty of nourishing milk from their mothers. But sometimes a newborn lamb or kid will suffer hypothermia because of inadequate mothering, a lack of regular feeding, or simply because the lamb overslept and forgot to eat in extremely cold weather. Hypothermia is the leading cause of pre-weaning lamb and kid goat losses in this country. Hypothermic lambs and kids will die if not attended to immediately. Many deaths can be prevented with a few simple tools and a basic understanding of how hypothermia kills.

Hypothermia is a condition in which the core body temperature drops, and the body's vital signs begin to weaken. Heart rate and respiration decrease, and the metabolism slows down. Past a certain point, the digestive system cannot help a lamb or kid overcome hypothermia. Without energy delivered

The Complete Granny Miller

properly and directly into the core of the body in the form of glucose, brain function is impaired, resulting in continuing weakness, confusion, drowsiness, coma, and the eventual death of the kid or lamb.

What follows is information you may need to know to save a little life. If you are a new shepherd or goat keeper, what I'm going to recommend may scare you. I encourage you to put your fears and apprehensions behind you. Do what you must do. Because if you don't, your hypothermic lamb or kid goat probably will die.

"Thermometers and a Keen Eye Will Save Lives"

That's no hype or exaggeration. Nothing takes the place of good observation in cold weather. A mildly hypothermic lamb or kid goat can often be found before things take a turn for the worse. Slightly hypothermic kid goats and lambs will commonly have a characteristic humped up look or will sometimes be found sleeping alone in a corner. Such lambs and kids can be fed a little extra and warmed up without too much risk.

The judicious use of a livestock rectal thermometer can save the life of most lambs and kids at risk for hyperthermia. If you don't already own a livestock thermometer, you need to get one. A human thermometer will work in a pinch.

When I take the temperature of a goat kid or lamb, I lay them across my lap. The thermometer is easily inserted with a little spit from me or Vaseline. I keep the thermometer in place for about 3 minutes. I've found it helpful to tie a piece of string or dental floss to the end of the thermometer so it doesn't get 'lost' while in service. I've never had this happen with a lamb, but it can happen with a large animal.

Normal Temperature for Lambs & Kids
- Normal body temperature in healthy lambs and kid goats is 102°F to 104°F
- Moderate hypothermia is 99°F to 102°F
- Severe hypothermia is below 99F° If you see this, your lamb or kid is in serious trouble.

Two Stages of Hypothermia Mean Two Different Treatments
When treating a lamb or kid for hypothermia, it is essential to determine the appropriate treatment.
- Lambs and kids **under** five hours old have a special type of internal body fat that will keep them safe for a few hours, depending upon the air temperature.
- Lambs and kids **older** than five hours have used up the supply of internal fat that they were born with. Those lambs **cannot** be treated the same way as lambs younger than five hours old.

If your lamb or kid has a body temperature of 99°F to 102°F and can still hold its head up, suck, and is **under 5 hours old**, warm sheep or goat milk is all you'll need. Approximately ½ cup of milk (120 cc) every 3 or 4 hours,

The Complete Granny Miller

administered by bottle or stomach tube, is suitable for a medium-sized breed lamb. Provide a little more milk for large-breed sheep. Should ewe or goat milk, or milk replacer, not be available, cow's milk (raw is best) will work in a hypothermic emergency. While cow's milk is not ideal for lambs or kids in a pinch, it will do. Melt about 1 tablespoon of butter for every ½ cup (120 cc) of whole milk or diluted canned evaporated milk. If you do not have butter on hand, find something else that is 100% animal fat: tallow, lard, chicken fat, bacon grease, or whatever: no Crisco, vegetable oil, or margarine.Ensure the milk is warm before feeding it. About 100°F to 105°F is perfect. An emergency, by its nature, usually comes without a warning. So, it's a good idea to have the proper supplies on hand before you need them.

Stomach Tubing
If your lamb is fully conscious and can hold its head up, but cannot or will not suck, and its temperature is between 99°F to 102°F, you should stomach tube the milk. Stomach tubing is an easy-to-learn skill and is a lifesaver for nearly all neonatal farm animals. A thermometer and a stomach tube used correctly will save more neonatal lambs, kid goats, and calves than any other thing I know of. I can't recommend it enough, especially if you live in a cold climate.
Using a stomach tube is quite simple. Here's how to do it:
You'll need milk, a jar or pan, and a large 60 cc syringe with a stomach tube or catheter. While sitting on a bale of hay or a bucket, lay the lamb across your lap or hold it between your legs while you stand. Do this well out of the sight of mamma sheep. Have the warm milk ready in the jar or pan. Remove the catheter from the syringe. Dip the end of the catheter tube into the warm milk to moisten and warm it up. Insert the tube into the corner of the lamb's mouth. Gently pass the tube all the way to the stomach. The distance varies, but it's about 7 to 11 inches in most breeds of sheep. If the catheter tube doesn't go in or only goes in a few inches, or the lamb starts to struggle, you are probably in the lungs and need to remove the tube and reinsert it. When a feeding tube is inserted correctly, the lamb will remain relaxed and will not struggle.
Draw up a full 60 cc syringe of warm milk. Place the syringe back onto the end of the catheter. Then slowly and gently depress the syringe to a count of 10. That's all there is to it.
To remove the catheter tube, pinch it tightly between your thumb and forefinger and remove it quickly. You don't want any drops of milk to aspirate into the lungs accidentally.

Intraperitoneal Glucose Injection
If the lamb or kid cannot hold its head up and its temp is 99°F or below, or if the lamb is unconscious, **do not** use a stomach tube. Do not bring the kid or lamb into the house or try to warm it up in any way. You could kill it. The lamb or kid can no longer produce energy through its digestive system. It must have glucose directly. Glucose should be given to the lamb or kid before warming, or the lamb or kid may die from hypoglycemia (low blood

The Complete Granny Miller

sugar).

An intraperitoneal glucose injection is an injection of glucose directly into the abdominal cavity. It is the best way to save the life of a lamb or kid older than 5 hours and with a body temperature of 99°F or less.

If you do not know how to give this injection, get your veterinarian to do it. Or have them teach you how to do it ahead of time. But in an emergency, if you can't get a veterinarian fast enough to help you, here's what you need to know to do it yourself. Be brave. It is scary the first time you do this on your own. Just remember, your lamb or kid is almost dead. You have nothing to lose and maybe everything to gain.

Here's How to Do It

The needle size **must** be a 1 x 18 or a 1 x 20 gauge used with a large syringe. A longer needle could nick internal organs, and a shorter needle will not reach the space in the peritoneal cavity. A 60 ml syringe works well. You need to use a sterile 20% glucose solution. You can dilute a 40% or 50% glucose solution with sterile water if necessary. The rough dosing is as follows:

- 50 ml for a large lamb or kid
- 40 ml for a medium lamb or kid
- 30 ml for small lamb, kid, or triplet

If using 50% glucose, boil the water first to sterilize it before mixing and diluting. For a large lamb or kid goat, draw up 20 ml of 50% glucose into a sterile syringe. Now draw up 30 mL of sterile water. The water can be quite warm. In fact, it works better if it is. In bitter cold weather, the warm water keeps the solution from getting too cold by the time you make it up and inject it into the lamb or kid. You want the glucose solution to be slightly above normal body temperature when it is injected. I keep the syringe warm by keeping it under my clothes and close to my body until I'm ready to use it. The injection site is located ½ inch to 1 inch to the side, and 1 inch below the umbilical cord stump. Have the warm syringe ready before you pick up the kid or lamb.

Hold the lamb or kid up by its forelegs in front of you while you lean against a wall or bales of hay. The lamb's back and spine are facing your front. By holding the kid or lamb in this way, the liver and other internal organs are dropped out of the way of the needle. The lamb or kid probably won't struggle much or at all. However, you need the lamb to be completely flaccid and relaxed before injecting the glucose. Wait a few minutes if need be for the kid or lamb to go limp. With an unconscious kid or lamb, this isn't an issue.

To give the injection, first steady yourself against a wall or bales of hay. Next, take the cap off the needle and insert the needle straight on and directly into the belly, aiming slightly towards the tail or butt. Slowly release the plunger on the syringe. That's all there is to it.

Now it is safe to warm the lamb or kid back up slowly. The lamb should receive a course of antibiotics for a week afterward. I use long-acting

The Complete Granny Miller

penicillin. But you should consult your veterinarian for his or her recommendations for the appropriate antibiotic in your area and the proper dosage.

Bottle Jaw

Bottle jaw is the vernacular term given to pendulous lower jaw swelling in sheep, cattle, and goats. The swelling is a soft tissue edema caused by anemia that is characteristic in animals that are carrying a heavy load of blood sucking internal parasites, better known as worms. Most often in sheep, the worms are *Haemonchus contortus*, often called 'barber pole worms' because of the red and white twisted appearance in large female worms. But other worms, namely *Ostertagia circumcincta* and *Trichostrongylus colubriformis,* can also cause a bottle jaw.

What you need to know about bottle jaw is that if you see it in your animals, you have a problem and must act quickly. Because, especially in sheep, most often the first sign of a heavy worm load is sudden death. In sheep, cattle, and goats, diarrhea, weakness, and weight loss will sometimes also accompany a bottle jaw. The treatment for the most common worms is simple.

First, check to see what wormers work in your local area. Internal parasites can and do develop local drug resistance.

Here's what I do:
Check all sheep with a FAMACHE card. A FAMACHE card is a color photographic card that matches a sheep or goat's interior eyelid color to the degree of anemia in an individual animal. It's a diagnostic tool to help livestock owners identify which animals require worming and which don't. Separate the animals into two or more groups. Administer the appropriate wormer to the treatment group. Confine the animals for 24 hours after worming to allow them to pass the dead worms. Ensure they have access to plenty of fresh water. Then release the animals onto clean ground or pasture that has been free from livestock for at least two months. Repeat in 10 days if necessary.

Not all wormers are the same. So be cautious and read the label carefully. Pay attention to the milk and meat withdrawal times if the animal will be used for food or is on its way to the sale barn.

The Complete Granny Miller

Supply List for Lambing & Kidding

The supply list that follows is pretty much interchangeable for both lambs and kids. Except for an old towel, a small jar of iodine, an ear tagger, and an Elastrator, I almost never have a need for anything else during lambing or kidding season. If you have more than just a few sheep or goats, it's a good idea to keep the things listed below just in case.

Because I promise you, trouble never picks a convenient time to come visiting. Both sheep and goats seem to choose the worst possible weather to have their babies. You can depend upon it.

Here's What You'll Need

- Faith. Most of the time, lambs and kids come into the world without any trouble. But sometimes things do go wrong. Begin every lambing season with the complete and unyielding conviction that God is the source of all life. Not you. Acknowledge human limitations and be prepared to accept both heartbreak and joy.
- Iodine and a small wide-mouth jar for dipping navels or disinfectant
- An Elastrator with bands for docking tails
- Ear tagger and tags for identification
- A livestock thermometer for detecting hypothermic lambs
- A feeding tube for hypothermic or rejected lambs
- Canned, powdered, or fresh goat's milk. Frozen or fresh sheep, goat, or cow colostrum for orphan lambs
- A bottle of 50% glucose or dextrose; a large 60 ml syringe and a few 20G x 1-inch needles for severely hypothermic lambs
- Injectable penicillin and 18G x 1 ½ inch needles for ewes that may need an antibiotic
- A clean plastic bucket for soap and water to assist a ewe with delivery if necessary
- Old towels for drying off lambs, hands, or carrying a slimy, wet newborn
- Butter, lard, or lubricant jelly for hypothermic lambs or for assisting a ewe with delivery

Pick the Time of Day When Lambs Are Born

Did you know that the way a pregnant ewe or doe is managed can influence when she will lamb or kid? It's true.

The time of day that a pregnant ewe or doe is fed can influence when she will lamb or kid and how quickly she'll do it.

Sheep and goats tend to lamb 6 hours prior to and 6 hours after the time of their grain or main feeding. Knowing this can be of great benefit to those who would prefer to avoid late-night or early-morning lambing and kidding.

The Complete Granny Miller

If you prefer lambing and kidding during daylight hours, feed your ewes and does around noon. It will result in most lambs and kids (70%) being born between 6 a.m. and 6 p.m.

Also, ewes and does tend to begin active labor when there is less human activity going on around them. Human presence can slow down active labor so much so, that I make it a habit to leave the barn when I see an animal in labor. Without me in the barn, a ewe or doe is able to get on with her business without the stress of the boss constantly watching over her.

Rejected Lambs

Every once in a while, I'll get a ewe that, for whatever reason, doesn't want to mother her lamb or lambs properly. In sheep, it is vital that within the first minutes after delivery, a ewe sees, smells, hears, tastes, and touches her newborn lamb(s). Those behaviors form a strong bond between ewe and lamb. It's worth mentioning that a ewe that experiences a relatively easy delivery sometimes will not have as strong a maternal instinct as the ewe that has had a hard labor and delivery. The intense pressure of the lamb in the birth canal immediately before the full delivery stimulates a ewe to accept her lamb. Sometimes, with an easy birth or twins coming quickly together, an inexperienced ewe will become confused. Often, she will accept one lamb at the expense of the other, and sometimes they both are poorly mothered. Any post-delivery interference between ewe and lamb can upset the natural course and cause the rejection of lambs or poor mothering behavior.

There are a few different tricks that I use to convince a ewe to accept her lamb(s) and get on with the business of being a mother. Each situation is different and requires an evaluation of all the circumstances. But the first and best trick I use, I call,

"Bring in the Dog"

Often, the presence of a dog will encourage a strong protective instinct in a newly delivered ewe. If a ewe gives me even a hint that she will not immediately accept her lamb or lambs, she is confined in a small area for a little old-fashioned convincing. Her lambs are safely moved out of the way, and a dog on a leash is brought in to within 3 to 6 feet of her. The dog is then tied to a post right in front of her.

A nervous ewe will sometimes try to escape. But most often, the ewe will either freeze in place or stomp her foot if there is no way out for her. Usually, she will not take her eyes off the dog. Her newborn lamb or lambs can then be safely brought back and placed next to her. Once she realizes she cannot escape the situation and that the dog is not going anywhere, she will usually either put herself between the dog and her lambs or ignore the lambs while watching the dog. This gives her lamb(s) a chance to root for a warm teat while she's preoccupied. Her lambs must get the first drink of colostrum milk from her. Ewes recognize their lambs by the smell of the

The Complete Granny Miller

lamb's backside. The lamb needs to smell like her.

Any medium-sized or larger dog will work. Sometimes the dog will have to be tied in front of her for 8 to 12 hours. For a truly recalcitrant ewe that will not allow her lambs to nurse, she is placed in a stanchion, or her back legs are hobbled and tied so her lambs can suckle. I leave enough room so that the ewe can get up and lie down comfortably. I'm not averse to stanchioning a ewe for up to 48 hours.

Good mothering in livestock is an inherited trait. I breed for it and suggest that you do too. On my farm, a first-time mother who rejects her lamb or kid is given the benefit of the doubt due to inexperience. She gets a reprieve and a second chance. But if in the following year she rejects her lamb again or proves to be a bad mother, she's loaded onto my truck for a trip to the sale barn as soon as the snow is melted off the lane.

How to Dock a Lamb's Tail

All lambs born on my farm get their tails docked for health reasons. Tail docking (shortening of the tail) is usually done in the United States to protect sheep against blowfly strike. Sometimes, when young lambs or adult sheep are on lush grass or are wormy, their manure becomes loose and coats their tails, rear ends, and back legs. When the weather is warm and humid, a wet, dung-coated sheep tail and backside is an attractive breeding ground for flies. It's the perfect environment for flies to lay their eggs. When those fly eggs hatch, they quickly turn into maggots that tunnel into the sheep's flesh. Because sheep are so wooly and fluffy, maggots can slowly eat the sheep alive without the shepherd being aware of the problem. Often, it's hard to tell what is happening until it is too late. Blowfly strike can be fatal. Tail docking does cause transient pain. However, experience has shown me that the benefits of tail docking far outweigh any temporary discomfort to the lambs.

Over the years, there has been much discussion about proper tail length. At one time, dreadfully short tails were the fashion in show rings across the U.S. Thankfully, nowadays most sheep producers will concede that no good comes from too short a tail. It is generally agreed upon that whatever method of docking is employed, the tail should be shortened to just beneath the caudal tail fold. A reasonable tail stump is long enough to allow the sheep to switch their tails to keep the flies from biting their tender parts. A sheep's tail should be long enough to cover the anus in rams and the vulva in ewes.

There are several different methods for tail docking, and the approach varies by location, culture, and tradition. Each has its particular advantages and drawbacks.The two most popular methods in the U.S. are banding and amputation.

Amputation is performed in various ways, and I believe it is the most prevalent method worldwide. Tail amputation is the preferred method to reduce the incidence of tetanus. With amputation, a lamb's tail is cut between the vertebrae with a sharp knife or scalpel. Care must be taken with older lambs, as excessive bleeding can sometimes be a problem. However, it can usually be controlled and stopped by applying a hot iron or a clean rag to the stump. Here in Western Pennsylvania, a few shepherds I know do use the amputation method. They heat bolt cutters until they glow red-hot and then cut the tail off. The heat from the bolt cutters cauterizes the tail stump, preventing bleeding.

The method I use and the one I think is simplest for beginners is the banding method. The banding method is completely bloodless. With banding, a heavy-duty rubber ring is placed over the tail with special pliers called an Elastrator. When the band is applied, blood circulation to the part of the tail beneath the band is cut off. The tail will wither and drop off on its own accord in about 10 to 14 days. The banding method carries with it the risk of tetanus. It's one of the reasons for ewes to be current with their tetanus booster shot before lambing. Neonatal lambs are protected from tetanus via the colostrum milk from their mothers. A tetanus antitoxin shot should be given to unprotected or at-risk lambs at the time of their banding. All lambs should be healthy and well started before having their tails docked. With banding, I think the ideal time is when a lamb is between 48 to 72 hours old. A lamb older than 7 days should not be banded, in my opinion.

When I dock tails, I hold the lamb between my knees and lift the tail. I then place the band on the tail, just below the caudal fold, and roll it off the elastrator with my thumb and forefinger. Sometimes in cold weather, the bands are stiff and hard to remove from the elastrator onto the tail. A little Vaseline applied to the Elastrator prongs can help make the bands slide off more easily. When the band is applied to the tail, there is pain for the lamb. But within 30 minutesor so, the pain appears to subside. Just remember that it's always kinder to leave the tail a little longer than shorter.

Lamb Creep

A lamb creep is a protected area away from adult sheep. It's a place where young lambs may sleep, rest, or play. The creep is usually designed with a vertical entry to permit young lambs to pass through, but not adult sheep. A low horizontal entry works well, too. Even an old tire will work. The point is to have an entry small enough so that adult sheep cannot get into the creep area.

A lamb creep is an important flock management tool and animal welfare consideration. When young lambs have access to a creep, they are able to get out of the way of adult sheep so that they aren't accidentally stepped on or trampled. A creep gives them a stress-free, warm, and dry environment away from the adults in the flock. A lamb creep has another vital function: feeding lambs. Grass is the most natural and economical food for lambs. Spring lambs grow well on good pasture. They often reach a market weight of 60 to 100 pounds, depending on the breed, by mid-to-late October. Some grass-fed autumn lambs are graded as USDA Choice. But most pasture-raised lambs are sold as feeder lambs. Feeder lambs are lambs that are

The Complete Granny Miller

finished on concentrates (grain, mineral, and legume mixtures) in feedlots to fatten them up before they are slaughtered.

There are instances when it is desirable to supplement what young nursing lambs are eating. In certain management situations, lambs benefit from a practice known as creep feeding. Creep feeding is the early introduction of concentrates into a lamb's diet. Nursing lambs that consume solids or concentrates are weaned sooner, make faster gains, and don't put as much of a drain on their mother's bodily reserves. This is a significant consideration in flocks that have a large number of multiple births. Or have older ewes that may not be milking adequately, or ewes kept on poor pasture. Lambs that are creep-fed can be marketed sooner and often get to bypass the feedlot altogether. I think anything that keeps an animal out of a feedlot situation is a good thing. By feeding lambs away from adult sheep, the lambs don't have to compete for food. The improved nutrition and weight gains appear to promote early maturity in ewe lambs, allowing them to be bred a bit sooner. Creep feeding tends to make weaning a little easier and more stress-free for the lambs. It also allows for earlier weaning. The disadvantage of creep feeding is that it is a more expensive way to raise lambs.

PIGS

Traditionally, pigs have been considered to be the most reliable means for low-cost meat for homesteaders and garden farmers. This is especially true for homesteaders with big families and many mouths to feed. Pigs, unlike other livestock, have a digestive system similar to humans. Pigs can and do thrive on an extensive and varied diet. They make good use of waste milk, kitchen scraps, and excess or rotten garden produce. Baby pigs thrive on goat's milk, and sometimes it can be cost-effective to buy and milk a dairy goat just for the benefit of feeding pigs.

Pigs are easily trained to electric fencing and can often be housed where other livestock cannot. But unlike sheep or goats, a full-grown pig can seriously hurt you. Thankfully, by the time feeder pigs reach the size and weight where they could injure you, they are ready for slaughter and won't be a problem for too much longer.

For small holders who need hard-packed bedding cleaned out of a building or barn, but don't have a tractor or a front-end loader, pigs are the answer. With a bit of encouragement and a small amount of grain, pigs can quickly turn hard-packed barn bedding into beautiful, ready-to-use compost. Because pigs naturally root forward, they are more than glad to help you get the barn cleaned out.

To encourage pigs to loosen up hard pack, simply open up a small, narrow hole in the hard pack bedding and pour a little grain into the hole, and turn the pigs loose. The pigs will naturally root forward into the bedding, loosening it up to get to the grain. Continue to pour a small amount of corn or feed in the direction you want them to move each day. And before you know it, the hard pack will be light and fluffy and ready for the garden.

When buying baby pigs, it is usually a good idea to wait until the middle of

The Complete Granny Miller

summer before purchasing them. One advantage of waiting is that baby pigs are usually less expensive after the 4-H kids buy theirs in the spring. Another advantage is that pigs make good use of garden waste and put on weight faster when the weather is warm and settled.

Feeding Kitchen Scraps and Garden Waste

Throughout most of the year, our ordinary food scraps and kitchen waste will end up in one of two compost piles we keep near the garden. But during the summer months, when we raise pigs, the situation changes. During those months, all the kitchen scraps and garden waste are fed to pigs instead of being composted. Pigs will eat most anything - including ham, bacon, and humans. Pigs make good use of kitchen and garden waste. In fact, we only raise pigs during the summer months when food is in abundance. Our garden always produces much more food than we can use. And if we should have excess goat or cow milk, so much the better. We feed it to the pigs. Nothing makes little pigs grow faster than fresh milk.

During the summer, my favorite thing to feed pigs is the foam that I skim off homemade jelly or jam. Our pigs squeal with delight whenever they see me coming their way carrying a bowl of jelly foam. The pigs like the foam because of the sugar. But the sugar in the foam makes them hyperactive for a couple of hours afterwards. I call it 'sugar rabies'. Watching a couple of pigs kick up their heels, run in continuous circles, and play in the mud, high on sugar, is my idea of country fun.

During the winter months, when we don't have pigs, our chickens will get the kitchen scraps that are too good for the compost heap. Chickens are omnivorous and will eat many things you wouldn't expect. Chickens love the rinds of cantaloupe and watermelon. Apple peelings are always a big hit, as is spaghetti or rice. Chickens love lunch meat and do like a little bit of raw meat now and then. In fact, if hens stop laying and aren't molting, sometimes feeding them a little raw ground meat for a few days will get them back on track. But chickens aren't like pigs. They can be fussy and do have food preferences that pigs don't have. Our chickens don't care for bananas and won't touch potato peelings. A pig would never let good food go to waste.

CATTLE

Cattle turn grass into beef. That's why cattle so often have a place on many small farms and homesteads. Frequently, it's the acquisition of cattle that begins the transition from garden farm or homestead to small family farm. Notice I used the plural - cattle. One family cow doesn't make a small farm. But with more than three or four head of cattle, the average garden farmer or homesteader could shortly be on their way to filling out an itemized

The Complete Granny Miller

ScheduleF 1040 come next April.

For most small holders, cattle represent a sizable financial investment, not just in the cattle themselves, but also in equipment, fencing, shelter, feed, and veterinary expenses. What's more, due to their sheer size and weight, cattle aren't as easily handled as pigs, sheep, or goats.

Cattle can hurt you, and hurt you badly. Every day in this country, farmers are injured or killed by their cattle. Most often, the mishaps aren't due to unsafe animals. But instead, they are due to a miscalculation, owner complacency, or a frightened animal.

Dexter Cattle

During the last twenty years or so, there's been an increased interest in smaller cattle for homesteads. At present, the most popular breed of diminutive cattle in the U.S. is the Dexter cattle. Dexter cattle are a minor breed and do have a place on many small farms and homesteads.

As a former Dexter breeder and owner, it always surprises people that I don't routinely recommend Dexters to homesteaders. There are many reasons for this. To begin with, there is no single, uniform type of Dexter. Many people don't realize it, but Dexter cattle come in a few dissimilar types. There is no one breed standard. Some Dexter breeders breed for a beef type animal. A few Dexter breeders are breeding for a dairy type. And unfortunately, some backyard Dexter breeders are breeding cattle for cuteness. Trust me on this. Cute is not a good reason to breed cattle. We're not talking about pet dogs or cats. We're talking about livestock.

The lack of a breed standard is exacerbated by the fact that, at present, there are three different Dexter breed registries in the U.S. The three registries have fostered a tremendous amount of infighting and back-biting among Dexter breeders and between the registries themselves. Many people who buy Dexters are either uninformed about the breed or lack general knowledge about cattle. When compared to other types of cattle, either beef or dairy, Dexters are expensive to purchase and are not budget-friendly for the average homesteader of modest means. Moreover, many breeders market Dexters as a feed-efficient animal that can be finished solely on grass. That's not exactly true. While it is true that some Dexter owners do keep their cattle on poor pasture, it is not true that beeves managed in such a way produce a meat carcass that is acceptable to most people. If you have poor or marginal pastures, Highland cattle are probably a better way to go. For their body weight, Dexter cattle eat more than you'd suppose and give less milk than you'd expect. Because of their smaller size, Dexters don't do as well at the sale barn as other cattle do. Dexter breeders are often forced to find a direct sale for their beef because of this market prejudice. That said, because Dexters are a small animal, they have a couple of distinct advantages for the homesteader or small holder.

For people with limited pasture space or meager pasture, a Dexter will do better than some other types of cattle. Because they are smaller, they are easier on facilities. Dexters, if treated properly, can be extremely friendly

The Complete Granny Miller

and affectionate. They are more easily handled than larger cattle. Of all the minor breeds of cattle, Dexters are the most mainstream. This is an advantage when buying Dexters and is also sometimes an advantage for breeding Dexters.

Due to their smaller size, a Dexter heifer or cow cannot be bred with just any breed of bull. A small-breed bull must service her. This is especially true for first-time heifers. Due to the popularity of Dexters, it isn't really too hard to find a Dexter bull. Also, Dexter semen is readily available should you decide to go the AI route for breeding. I recommend AI for people who are just keeping a few cows.

If you decide that Dexters are a breed of cattle that you'd like to keep, just keep in mind that there are a couple of genetic traits that you'll want to avoid like the plague. The first and most serious fault is known as a 'bulldog calf'. A bulldog calf is actually called chondrodysplasia. It is a lethal genetic mutation and is form of dwarfism. Another genetic mutation is known as a 'water baby'. This is the folk term for pulmonary hypoplasia with anasarca (PHA). PHA in cattle is a condition characterized by incomplete lung formation. The fetus or calf has a monstrous, swollen, and bloated appearance due to water retention. Such calves often weigh twice what they should. Many such calves are never born alive. Unfortunately, PHA can cause severe calving difficulties. Both conditions are significant problems in some strains of Dexters. Thankfully, both syndromes are being limited today by genetic testing. Dexters, like all cattle, will benefit from an intelligent and well-thought-out breeding program.

Dexter Cattle Flaw – Bad Feet & Udders

One pet peeve that I have with Dexter cattle is their feet. A neighbor who runs cattle on nearly 1000 acres stopped by the farm the other day. We were standing near a pasture with some Dexters, and he remarked that one of our little Dexter steers had the worst feet he'd ever seen on an animal so young. I laughed and replied, "You just ain't seen enough Dexters."

And that's true enough. Because it's hard to believe that one breed of cattle, heritage or otherwise, can have so many faults until you actually see lots of them together in one place from many different breeders.

In June 2012, my husband got a chance to do just. He attended the National Dexter Cattle Show and Sale in Fort Wayne, Indiana, to meet with other Dexter breeders. He was appalled by what he saw there. To this day, he still calls the 3-day affair 'The Bad Feet and Bad Udder Show.'

The faults of both poor feet and udders are a plague upon most strains of Dexter cattle. Both faults can be corrected through intensive culling and an intelligent breeding program. But sadly, few Dexter owners are willing to ruthlessly cull inferior animals due to ignorance and financial considerations.

Many Dexter cattle are sold to new and not-so-new homesteaders who want a smaller 'dual purpose' animal, but have little experience with dairy or beef cattle breeding and husbandry. Folks, there are no satisfactory dual-purpose cattle. Dual-purpose doesn't do much of either purpose very well.

Inexperienced people simply don't know that, and backyard breeders are doing precious little to improve the Dexter breed as a whole. The truth is, for most people looking for a family milk cow, Dexters are way overpriced, eat too much for their size, and are frankly overhyped. As beef cattle, they still eat too much and take a long time to bring to slaughter weight, especially if you are buying feed. Don't let the small size fool you. Dexter cattle are not feed-efficient animals, nor are they cattle for people of modest means or families on a budget.

Now I know my opinion hasn't made me popular with other Dexter breeders. And it won't make me popular with homesteaders who've always dreamed of owning a so-called dual-purpose Dexter for milk and beef. But don't get me wrong about Dexter cattle. I'm simply the messenger. Dexters have a real place on certain homesteads and garden farms. But do not believe everything you read about them on the Internet.

A Realistic Assessment of Kerry Cattle

My husband and I originally bought a herd of Kerry cattle for an integrated dairy production project on two different family farms. We wanted to introduce the genes for longevity, feed efficiency, and A2 beta-casein into an existing dairy herd. The plan was that our neighbors would keep them and milk them along with their dairy cows on their farm. We would raise the heifers, and feed out the steers for beef on our farm. The plan did not work out because of the extremely low milk production. The Kerries weren't earning their keep. They gave less milk than a good dairy goat. Lately, I have been trying to figure out the practical place Kerry cows might have on my farm and in 21st-century America.

The Good

- Kerry cattle are small but are not dwarfs. They are not nearly as small as many Dexters. The larger size is an advantage for those who lack a reliable means to direct sell to consumers and must sell feeders or finished beef through traditional livestock markets.
- Excellent feed efficiency. Kerry Cattle eat about 25% to 30% as much feed as a Holstein and maybe 50% to 70% as much feed as a modern commercial Angus.
- Kerry cattle are tame if handled right
- In crossing with other breeds, the solid black color is dominant. Black cattle can bring a premium in traditional livestock markets. Black cattle fetch 12% to 30% more than other colors of cattle.
- Extreme longevity. There are verified instances of 20-year-old Kerry cows having calves
- Their primitive DNA may provide some unique disease immunity. There was initial research in Britain indicating that no Dexter or Kerry cow had ever been slaughtered for BSE. But this was never thoroughly researched.

The Complete Granny Miller

- Beef flavor was traditionally well regarded. My experience with Kerry beef confirms this.
- Small birth weights. A smaller calf means less calving assistance by the herdsman in order for a calf to be born.
- High rates of A2 beta-casein. A2 beta-casein is a source of controversy in some dairy circles at present. Most dairy cattle in the US do not carry the gene for A2 beta-casein.
- Kerries have a distant relationship to other breeds of cattle. This maximizes the advantage of heterosis in crossing with more mainstream breeds.

The Bad

- Most Kerry cattle in North America today come from one single importation from Ireland to Canada. The gene pool in North America is extremely limited. All Kerry cattle in North America are closely related.
- From at least 1919 onward, the majority of Kerry breeders in Ireland and England were aristocrats. The landed gentry kept Kerry cows as prestigious estate cows. Kerry cattle have not really been bred for milk production or improvement for at least 90 years. This is a serious problem today, as most North American Kerry cattle breeders keep them for their rarity or heritage, and not for milk production.
- My direct experience with milk production was an abysmal 15 pounds per day. That's less than 2 gallons a day over a 180-day lactation period. Modern dairy cows have a lactation of at least 305 days. Even for many home dairies or people keeping a family cow, this is just not enough milk. It's enough milk for a calf and about half a gallon a day for the table or kitchen. It is undoubtedly not enough milk to make getting pooped on and occasionally kicked worthwhile.
- Kerry cattle are a genetic dead end. There is no consistency in owner expectations about the future niche of these cows.
- There is not enough modern data on crossbred performance. Crossbred performance is the heart of modern-day beef production and is becoming common in dairy cattle breeding. Many of the 100-year-old reports on crossbreeding Kerry outcomes are no longer relevant today. This is because every other breed of cattle that has been crossed onto a Kerry has changed since then.

Head Whorls in Cattle

Selecting young cattle for future breeding is as much an art as it is a science. It's a guessing game regarding future productivity, milking ability, and mothering. When selecting foundation breeding stock, good sound feet, sturdy legs, a nice straight top line, and a feminine appearance are all important considerations.

The Complete Granny Miller

Temperament is also an important trait, especially with a small herd. Like many other animals, temperament in cattle is hereditary. The truth is, some cows are just born wild, and there's not much to be about it. Such cows are harder to handle and can end up hurting themselves or you. Crazy cows will run into walls and fences trying to get away from people. Crazy cows can be dangerous in a squeeze chute and in an open pasture. Life is too short to put up with crazy cows.

When selecting cattle for purchase from a large group, I judge cattle temperament on 3 factors:

- How the cow or heifer reacts to strangers in the lot
- Whether or not the cattle 'crash' during handling
- The position of head whorls

The first two selection factors are obvious. Head whorls are maybe not so obvious.

A facial head whorl is where the hair on a cow or horse's face meets. Head whorls are simply cowlicks on the face. Generally, the lower the whorl is on the face, the less high-strung and calmer the animal will be. The head whorl should be at or just above eye level on the face.

It is best to avoid high whorls and especially uneven whorls high on the face. Selection by head whorls is not foolproof. But when combined with the other two factors, it can help pick good heifers out of a large lot.

Horns on Cattle

The first or second thing most farm visitors notice about our older Kerry cows is their horns. It's understandable. Horns were what I initially noticed about Kerry cattle the first time I saw them, too. From a farmer's point of view, horns in livestock can be a source of trouble. Horned cattle are more dangerous to handle and are subject to accidents. When we bought our first Kerry cows a few years ago, they already had large horns. For the most part, our cows are good girls. They don't misbehave too badly or get into trouble with their horns. However, we made the decision early on in our Kerry breeding program to de-horn all younger heifers and steers that were born on this farm. It makes life easier for both humans and cattle.

The presence of horns in cattle is the result of genetics. In cattle, the genetics for horns are recessive. Cattle that lack specific genes are naturally hornless and are known as 'polled'. Some breeds of cattle, like the Angus and Galloway, are always polled. In other words, hornless. The gene for polling is dominant in cattle. In this area of the country, old-timers used to refer to a naturally polled cow as a 'Muley Cow'. I always wondered why they were called that. Then I learned about the naturally polled Moiled

The Complete Granny Miller

Cattle that once roamed Northern Ireland. Moiled and Muley sound a lot alike, and I believe the word may have survived from the first Scots-Irish settlers in this area.

Before we bought our first Kerry cows, the previous owner had allowed a commercial crossbreed Simmental-Angus bull to breed two of them. The resulting offspring were born naturally polled. I like to think of them as 'Improved Kerry Cattle'.

Cattle with horns may have a place in an extensive livestock raising situation. Cattle that spend a lot of time in semi-wild places can use their horns to knock over young trees and brush for food, and as a defense from predators. However, in situations where cows are managed inside buildings and yards, such as during our long Pennsylvania winters, horns can cause small puncture wounds on other cattle and even their keepers. Horns can catch on pipelines, gates, and feeders. And a broken horn on an adult animal can be an absolute bloody mess.

But horns do have practical purposes. Before plastics, horns were used for buttons, cups, powder horns, and other items. Once removed from a slaughtered cow, horn material can even be heated and shaped. Horns on cattle are especially useful in working oxen. The horns keep them from backing out of a yoke. Horns once served as a defense mechanism for cattle. I find it curious that a genetic factor, which would make cattle less able to defend themselves in the wild and make bulls less able to beat rivals to build a harem, would be dominant. Regardless of the gene expression, horns on cattle remain both a source of expense and controversy. In the dairy cattle world, all mainstream dairy breeds still have horn genetics. Horns cost the dairyman time and trouble to remove. With cattle, the de-horning procedure should be done as early as possible in a young bull or heifer's life to avoid unnecessary pain and the possibility of complications. But de-horning must be done with regard to weather conditions. When improperly done, de-horning can result in sinus infections and fly strike, which can harm the newly de-horned animal.

Polled dairy genetics are now available. But most dairymen have not used them. Some farmers are concerned that breeding for a single characteristic, such as polling, could result in the loss of other valuable genetic traits. In the Dexter cattle world, naturally polled genetics seems to be a matter of some controversy. In fact, it was the cause for the creation of one of the three Dexter registries.

Fly Control in Pastured Cattle

Here in Western Pennsylvania, summertime means fly season for folks who raise cattle. Because cattle produce a significant amount of manure, fly populations can be challenging to control, even when cattle are kept in open pastures and not confined. Heavy fly infestations in cattle can be a real animal welfare headache. That's because flies bite, suck blood, spread disease, and cause agitation and general misery, both in cattle and humans alike.

The Complete Granny Miller

In this part of the US, there are two major types of flies that trouble cattle:

- Face Flies
- Horn Flies

If you are new to keeping a family cow or plan to keep cattle in the future, it is a good idea to learn about the differences.

Face Flies
Face flies look a lot like big house flies. They tend to cover large areas of the face and like to feed on the eye, nose, and mouth secretions that cattle produce. Face flies transmit the bacteria *Moraxella bovis*, which is the primary cause of bovine pinkeye. Pinkeye is a highly contagious inflammation of the cornea and conjunctiva of cattle and can result in blindness if not treated promptly.

Horn Flies
Horn flies are about half the size of house flies and have pointier-looking wings. As the name suggests, they prefer to gather in a large mass around the horns or poll of cattle. But horn flies don't just stop with the head. They are blood suckers and congregate wherever they can't easily be rubbed or brushed off. Horn flies like to stay on cattle continuously. They will often face directly downward towards the ground while they cling to cattle. The back and shoulders of calves and full-grown cows are all easy targets for horn flies.
During rainy weather, they will move to the belly and throat of cattle for shelter from the rain.

Fly Control
There are a few different types of products and delivery systems that aid in the control of flies on cattle. The type of fly control that a producer or smallholder will choose depends on their budget, herd size, management system, and personal preference.
Spray On
A daily insecticide spray is a relatively effective control measure if you have only a few cows and handle them daily. It's the type of fly control I use, and it provides about half a day's protection.
Pour On
A pour-on fly repellent is a good choice for both dairy and beef cattle and

The Complete Granny Miller

can last up to four weeks. I keep pestering my husband to get some so we can quit with the daily spray.

Dust Bags or Oil Rubs

Dust bags and oil rubs are convenient and can be placed between two fence posts or a gate where cattle will walk daily. Dust bags are filled with powdered insecticide that is applied like bath powder on bovine heads, backs, and necks when they walk underneath it.

Oil rubs are highly effective and resemble the big horizontal wiper mops commonly found in automatic car washes. Instead of a car passing through and being doused with soap and water, a cow passes through and gets doused with insecticide.

Ear Tags

Insecticide ear tags are an effective but expensive form of fly control. Best results are obtained if tags are set in both ears and not too soon after the start of fly season. Caution and rubber gloves must be used when applying insecticide ear tags. Flies can develop a resistance to the active chemicals after a couple of years of use.

Mineral or Feed Additive

The way a feed or mineral additive works is that cattle eat the insecticide in the feed or mineral block, and it passes through the digestive tract and into the manure. The insecticide reduces the number of flies emerging from the manure, helping to keep fly populations under control.

Freemartin Heifers

A freemartin is an infertile female mammal. In cattle, a freemartin is the normal outcome of mixed sexed twins, with 90% - 95% of all heifers born sterile.

Sometimes, a single heifer calf will be made sterile from the death of a male twin during the early part of the gestation period. The female twin is made infertile in utero due to an interconnection and fusion of chorions and shared blood vessels, which permits the blood from each twin to flow around the other twin. The action of male hormones upon the female fetus usually renders the heifer calf with non-functioning ovaries and a short vagina.

The heifer often displays a masculine appearance and behavior. The action of the female hormones on the male fetus usually does not affect the bull calf, except sometimes the testicles may be smaller than usual. This is important because testicle size is associated with cattle fertility.

The freemartin effect has been recorded in goats, sheep, and pigs, but it is rare. The 18th-century Scottish physician John Hunter was the first to observe that a freemartin heifer always has a male twin. Freemartinism is usually diagnosed by vaginal examination with a probe at 3 to 6 months of age. Vaginal length in a normal heifer calf is usually greater than 5 ½ inches. In a freemartin heifer, vaginal length is usually between 2½ to 4 inches. Sometimes, genetic testing is used with a valuable heifer to determine breeding ability.

The Complete Granny Miller

But most often, a visual examination of the placental membranes shortly after birth will confirm the probability of sterility of the female calf.

The Castration of Bull Calves

Dairy bull calves make a good backyard beef project. They can often be had for a reasonable price. If you don't mind waiting 18 months before the beef is on the table or in the freezer, it may be just the thing for your family. Under normal circumstances, calves raised for beef are castrated not too long after their testicles descend. Sometimes, a producer will wait to castrate if there is a possibility that a particular bull calf may be worth saving as a breeder.

Castration is any procedure that results in a male animal losing the function of the testes. When a bull is castrated, he is known as a 'steer'. The castration of farm livestock is performed for a few different reasons. The primary reasons are that castration yields better meat quality and a more marketable carcass. Castration lessens the aggressiveness in male animals and prevents reproduction and sexual behavior. Circulating testosterone in male animals affects the meat quality. Meat from intact male animals, whether it's a rooster, a ram, or a bull, is tough, stringy, and strong-tasting. Because castration significantly reduces aggressiveness in livestock, it results in less wear and tear on fences and facilities. A calmer and more docile animal is safer for humans to handle and is safer to other animals. Bulls, especially dairybreed bulls, can be exceptionally dangerous as they grow. Never trust a bull. They can be deadly.

In general, there are 3 methods of castration used with livestock:

The Surgical Method
This method of castration is the most direct and sure. The surgical method of castration involves cutting the bottom of the scrotal sac and removing the testes. The advantage is that the castration is complete. There is never a doubt because both testes end up on the ground or in your hand. The disadvantage of this method is that it involves cutting, which runs the risk of bleeding or infection. With this method, I use an almanac to help pick the best day.

The Elastrator Method
This method is known as banding and is bloodless. A small rubber ring is applied around the scrotum with a special type of pliers known as an Elastrator. Banding should be done only at a sufficient height on the scrotum to accommodate both testes within the band, and no higher. When the rubber ring is applied, the blood supply is interrupted. The testes and scrotum atrophy, wither, and eventually fall off. It is a good choice for young animals if they or their mothers have been vaccinated for tetanus. The drawback is that if they are not protected against tetanus, they run the

The Complete Granny Miller

risk of acquiring tetanus and may die. Tetanus in livestock is almost always fatal. Another disadvantage to the banding method is that if one or a part of a testicle is missed, testosterone is still produced, resulting in a 'staggy steer'. A staggy steer shows many of the same physical traits as a bull: a muscular neck, a broader crest, and head. This is not desirable. Staggy steers bring much less at auction.

The Emasculatome Method

This method is sometimes known as the Burdizzo method, and it is also bloodless. The Emasculatome method uses a special instrument that crushes and rapidly cuts off the blood supply to the testicles one at a time, via the spermatic cords. This causes the testicles to soften and shrink, and then be reabsorbed by the body. The advantage of the Burdizzo method is that it poses little to no risk of tetanus and is a quick procedure. The disadvantage is that, like the banding method, if not done correctly, it will result in a staggy steer. The advantage is that it can be used during fly season because it is a bloodless method.

Most beef cattle producers in my area of the country use the Elastrator method and band beef calves while they are young. That's because beef cattle grow faster than dairy cattle and are usually spookier with people. With bottle-raised dairy bulls, it may sometimes be beneficial to delay for 2 months or so to promote faster and earlier growth, and to minimize any setbacks to the calf. Castration by cutting is not painless. Some veterinarians and cattlemen may choose to administer a local anesthetic, while others may not.

The Care of Umbilical Cords

The vast majority of farm livestock are born healthy, strong, and without any human intervention. For the most part, nature does a pretty good job when given a fair chance. However, it is essential to be realistic about animal life. In some livestock situations, health problems and heartbreak cannot be avoided.There is no such thing as a 100% guarantee when it comes to living things. That said, many livestock health problems can be avoided with good and conscientious animal husbandry. One problem that is more or less preventable is something called 'navel-ill' or'joint-ill'.

Joint-ill is a serious bacterial infection. It is a common cause of lameness in neonatal foals, calves, lambs, pigs, and kid goats. The infection enters the body by way of the navel. Navel-ill is usually caused by strains of *E. coli* and *Streptococcus*. These bacteria tend to thrive in damp, dirty bedding and muddy yards. Navel-ill is a serious condition that requires prompt medical treatment with long-acting antibiotics. Navel-ill can be fatal if not caught in time. Baby animals that do recover from navel-ill often fail to thrive or will sustain permanent damage to their joints, eyes, heart, brain, or other organs. The symptoms of navel-ill are:

- Joint stiffness

The Complete Granny Miller

- Warm or hot joints
- Redness or pus around the navel
- Loss of appetite
- Fever

Like many livestock situations, prevention is the key to avoiding trouble. Animals are born with wet and sometimes bloody umbilical cords. The umbilical cord makes a perfect wick for bacteria to enter the body when it's dragged along the ground, or when the baby animal is lying on dirty bedding. Ensuring that all animals give birth in a clean, dry environment will significantly reduce the risk of navel-ill.

The application of an iodine navel dip or other disinfectant to the umbilical cord and navel area as soon as possible after birth will reduce the risk of bacteria entering the body via the navel. The iodine disinfects and dries up the cord and navel area. I think an iodine dip works best for small animals, and I prefer it over iodine spray.

If I can lift the newborn animal, I apply iodine directly to the navel area. Spray bottle iodine can be used for larger animal babies, such as calves or foals. The important thing is that the area be well saturated with iodine or disinfectant. I don't hesitate to reapply iodine if the cord doesn't seem to be drying up within a day or so.

Also, it is imperative that all newborn animals receive colostrum milk within 6 hours of birth. Colostrum milk is rich in maternal antibodies and helps protect young animals against disease. Good animal management practices, sanitation, and old-fashioned common sense will go a long way in helping newborn livestock get off to a good start.

The Complete Granny Miller

ANIMAL WELFARE

In 2015, I was fortunate enough to have a young veterinarian write
a few guest posts for GRANNY MILLER.

Dr. H. graduated in 2013 from Ohio State University College of
Veterinary Medicine. Raised on a small homestead here in
Western Pennsylvania, Dr. H. is well versed in the needs and
concerns of small farmers and homesteaders, and the animals
entrusted to their care.

How to Give Animal Injections

As a veterinarian, I'm always giving animals injections for one reason or
another. I give vaccines, I give antibiotics, and I give pain medications and
other drugs. Often when I'm treating an animal, I give owners instructions
on how to administer the medication themselves so that they can continue
the treatment. Many of my clients already know how to give injections, but
sometimes I have to teach folks how to give injections. Depending on the
medication I'm administering to the animal, the route of injection may vary.
What follows are the two different types of injections I routinely have
owners give to their animals.

- SQ or Sub Q – This is short for subcutaneously. This means
  you give the shot under the skin.
- IM – This is short for intramuscular. This means a shot is
  given into the muscle.

There are also IV (intravenous) injections. But I do not teach owners how to
do these types. Intravenous injections require a higher skill level. IV
injections also carry a higher risk of complications, especially if something
goes wrong. Some drugs or medications will cause abscesses and extensive
tissue damage if they leak out of the vein. If you should miss the vein and
give a medication in the artery, the animal will go into seizures.
When giving SQ or IM injections, you must always draw back on the
plunger to make sure you are not in a blood vessel and that you are not
drawing air into the syringe. If you do get air into the syringe, it means
you've generally gone through the skin and back out again. You don't have
to draw the plunger back very far, just enough to know where you are.

Where to Give an Injection

Where to inject medication depends upon the species of animal and the
preferred route for drug delivery. Food animals tend to have injection
guidelines based on meat and hide quality. Some medications can cause lots
of muscle scarring, which has to be later cut out of the carcass. Injections
also may cause scarring in the skin, which can devalue an animal's hide.
Areas to inject are chosen for the ability of that particular spot to handle an

---

The Complete Granny Miller

injection.

The SQ space in some animals can handle quite a bit of drug volume. But muscles, on the other hand, can't take as large an injection. Muscle injection sites are chosen based on muscle size, animal comfort after the injection, and consideration of the muscle value as a cut of meat. Muscles should not be given more than 10 ccs (mLs) of medication at one time. Large volumes of medication for muscles need to be split into two or more injection sites. What follows are the common farm animal species and where to give them injections.

Horses

Many injections given to horses are IM. These injections are given either in the neck or, less commonly, in the muscles farthest towards the hind end. The neck is most often used because it is also safer for the person giving the injection, as they're less likely to be kicked.

Small Ruminants
Sheep and Goats

Many injections given to small ruminants are SQ, but sometimes they are given IM. Small ruminants are considered food animals even though many people have them as pets. The most common injection site for sheep or goats is right in front of the shoulder. This shoulder spot has loose skin, which is suitable for SQ injections, as well as muscles for IM injections. Another spot for SQ injections in small ruminants is behind the elbow, where the hair or fleece is thin. At this site, it's easier to find skin.

Cattle

Cattle are popular food animals. Even when a cow starts in a dairy herd, she often ends up going to slaughter when her milk-producing days are over. Because of this, almost all cattle injections are given in the neck region. The neck region on cattle is a low-value meat area and is not a valuable part of the hide. Both SQ and IM shots are given in this area. The one shot that is given differently is the antibiotic Excede, which is given SQ at the base of the ear.

Swine

Pigs and hogs are another popular food animal, which some people also keep as pets. The ideal spot to inject a pig is in the neck, about 7 cm behind and below the base of the ear. Again, the area chosen for injections for pigs was decided based on meat quality and the ability of the area to handle injections.

Camelids

Llamas and alpacas have gone up and down in popularity, but are still quite numerous. Many folks have them as guardian animals or for fiber. Because camelids are fiber animals, injections can be a little interesting. There are a couple of places to give SQ injections. One is in front of the shoulder, and the other is behind the elbow, down where the thick fleece ends. Most

The Complete Granny Miller

injections in camelids are SQ. But should you have to give an IM injection, the recommended spot is in the muscles of the shoulder, above the elbow.

No matter what medication you are giving or where you are giving it, there are some things you should always do for cleanliness and animal safety. Always use alcohol to clean off the top of the bottle. Use alcohol to clean a spot on the animal where you are giving the injection. This helps keep your medicines or vaccines uncontaminated, and it helps keep the needle from dragging dirt into the bottle or the animal.

Always use new needles going into a bottle (prevents contamination), and you should really only use one needle per animal. Sharp needles don't hurt as much, and you won't carry disease (like blood parasites) from one animal to the next.

Throw out contaminated bottles. This one can hurt, depending on the medication or vaccine. But what hurts worse? Possibly injecting bacteria or fungus into your animals and causing abscesses or systemic disease, or losing a few bucks?

So those are the main points about giving injections and where to give them. As always, if you have any questions, ask your local vet for help.

Dr. H

Supplies to Keep on Hand for an Animal Health Emergency

Animals rarely pick a convenient time to have an emergency. So, when one comes up, it is a good idea to be prepared with some basic supplies on hand and to keep a few things in mind. For starters, ask yourself: Is this animal a pet, or is it intended for the food supply?

If it is a food animal that is injured, is this animal ready to be butchered? Or can you afford to wait until the animal is healed up and the drugs are out of its system? For food animals, you have to be very careful of what you give them and the dosing. Because the drug withdrawal times for meat and milk are set to keep people safe, once these questions are answered, you can act accordingly.

General Supplies

Some general things are good to have on hand, no matter what species you have.

- Disinfectants - Betadine and Chlorhexadine are examples of disinfectants.These types of disinfectants are diluted with water. With them, you can clean off injuries and flush wounds and abscesses.
- Bandage Materials - Gauze rolls, gauze pads, vet wrap, cotton leg wraps, roll cotton, white bandage tape, and even in a pinch...duct tape. All of these bandage materials should be on hand in case of injuries. Even for an injury that you will need a vet for (like a

The Complete Granny Miller

broken bone), you can help your animal by cleaning off the injury or stabilizing it until the vet arrives, or before you transport your animal to the vet. Broken legs can benefit from lots of cotton padding secured with vet wrap. Gushing blood can be slowed or stopped by the application of pressure from a wad of cotton secured by vet wrap.

- Blood Stop Powder - You put it on to help clot blood. Flour and corn starch will also work.
- Electrolytes - Animals who don't want to drink or have diarrhea and vomiting, or who are ill in general, could all use some electrolytes. These replace the ones they are losing to diarrhea or vomiting, or aren't taking in.
- Drenching Gun or Oral Syringe - For those animals who are eating or drinking, or for those who may need oral medications.
- Gloves - Short exam gloves and the long rectal sleeves. The short gloves are ideal for handling injuries or administering oral medications, helping to keep your hands clean and protected from zoonotic diseases. They also help protect your animal's injury from the bugs that normally live on your skin. Rectal sleeves are good to have on hand when you are birthing animals and have to go in to reposition a baby. Again, they keep you clean and also keep the animal protected from your skin flora.
- Wound treatments - for minor wounds, things like Scarlet Oil and general triple antibiotic ointments are useful to have on hand. Furazone ointment is often used in horses, but is a major DON'T in food animals. AluShield is a spray-on bandage that is useful for covering minor wounds to keep the flies off.

Medications

This is where you have to start thinking about the animal you are treating and the purpose of its life. For food animals, you have to pay attention to medication withdrawal times and to which medications are forbidden. Your vet can help direct you, because what is good in one species can kill another. These are a few medications your vet may want you to have on hand.

Antibiotics - For most dogs and cats, your local vet can dispense or prescribe these according to diagnosis.

For food animals (cows, goats, sheep, pigs), many of these medications are commonly found at your local feed store. The type of medication your vet advises will depend on why you are treating your animal. Penicillin G is an oldie, but a goodie. It's a broad-spectrum antibiotic that is the first line of defense for many ailments, from wounds to retained placenta. Oxytetracycline-aka- Biomycin or LA-200. This is another broad-spectrum antibiotic that is used in a plethora of ailments, from foot rot to pneumonia. Ceftiofur is also a broad-spectrum antibiotic. These include Excenel, Excede, Naxcal,and Ceftiflex. Again, these have a wide range of uses. For horses, Penicillin is also commonly used, but SMZ-TMP is another

The Complete Granny Miller

broad-spectrum antibiotic your vet may want you to have on hand. Other antibiotics are used to treat a specific ailment and may only be prescribed when needed.

Pain Meds and Anti-inflammatories - These are most commonly prescribed as needed for pain. Often, people want to use human medicines, like ibuprofen, Tylenol, and aspirin, for their animals. But animals metabolize these drugs differently than we do, and giving them our medicines can kill them. So please, before giving your animal something, ask your vet.

For large animals, Flunixin Meglumine (also known as Banamine, Prevail, Flunixiject) is often one that a vet will let you keep on hand if there is a need for it. For large animals, it's their version of an NSAID. This medication comes in an injectable form for food animals and horses. But for horses, there is also a paste version for owners who are uncomfortable with giving injections. Horse owners may also want to keep Phenylbutazone (aka- Bute) on hand.

Here are a few other odds and ends that you can have on hand as well.

- Activated Charcoal is frequently used when an animal has eaten something toxic, like a poisonous plant.
- Mineral oil is often a go-to for colicking horses. But it is also used when an animal has eaten something you want to have move smoothly through the guts with some lubrication.
- Baking Soda is often used in ruminants to treat rumen acidosis
- Diphenhydramine is primarily used in pet animals. It isn't labelled for food animals, but it is effective in treating allergic reactions.

Vitamin B Complex and Thiamine are helpful for ruminants that are off their feed or ill. Ruminants usually make their B vitamins in their rumen, but when they don't feel good, they don't always make enough. Vitamin B Complex and Thiamine also stimulate appetite to some degree.

There are many other things you could have on hand for animal emergencies. These suggestions are just a starting point. Each farm and home is different. What one needs on Farm A isn't necessarily what will be needed on Farm B.

In any case, talk with your local vet ahead of time about the things you may need in an emergency. Your veterinarian can help you tailor your emergency grab bag to fit your farm's needs.

Dr. H

173

Common Veterinary Abbreviations

With every profession, there seems to be a secret language or code spoken for the purpose of confusing everyone else. Veterinary medicine is no different, with our alphabet soup of acronyms and abbreviations that we use as shorthand. Usually, I remember to ask my clients if they understand what I mean when I give them instructions. But sometimes I forget. So here are some of the ones I commonly use in practice and what they mean.

When Giving Injections
SQ or Sub Q - subcutaneously, this means you give the shot under the skin. Most animals, you can tent the skin and slide the needle into the tent.
IM - intramuscular, this one is a shot given in the muscle. For most animals, you give the shot in the neck muscles. For some, the shot is given in the muscles of the back leg. I'll often show my clients how and where to give these shots before making them do it on their own.
IV - intravenously, this is an injection I will not have you do because so many things can go wrong if done improperly. This is an injection into a vein.With most farm animals, I use the jugular vein; with dogs and cats, I use their leg veins.
IN - intranasally, meaning into the nose. Equine strangles vaccine and canine kennel cough vaccines are often given this way.

When Giving Medications
- **PO** - per os, this simply means giving something by mouth or oral medications
- **SID** or q24hrs - this is shorthand for giving something only once a day
- **BID** or q12hrs - shorthand for giving something twice daily
- **TID** or q8hrs - shorthand for giving something three times daily
- **QID** or q6hrs - shorthand for giving something four times daily
- **EOD** - short for giving something Every Other Day
- **mL or mil or cc** - milliliter, aka cubic centimeter, both mean the same amount. 1 mL is equal to 1 cc. This is the amount of liquid I want you to administer as an injection or orally.
- **mg or mig** - milligram, this is an amount and a weight
- **kg or kig** - kilogram, this is a unit of weight that the rest of the world uses, and so it is the standardized way for veterinarians to record and calculate weights. One kg is equal to about 2.2 pounds or lbs.
- **mg/mL or mig per mL** - milligram per milliliter, this is a concentration. For example, the common antibiotic LA-200 has a concentration of oxytetracycline of 200 mg/mL. The anti-inflammatory Banamine has a concentration of 50 mg/mL. I use these concentrations to determine how many mLs I need to give to your animal.
- **mg/kg or mig/kig** - milligram per kilogram. When I'm calculating doses, I'm often given the appropriate dose as "give this number of

mg per kg of animal weight." For example, with Safeguard, the dose for goats is 10mg/kg. I need to know your goat's weight in kilograms so I can calculate the correct dose by combining multiple factors.

- **IU** - international units, this is another concentration often found on the labels of penicillin and vitamins. Instead of mg/mL, the concentration is IU/mL
- **X** - I use this one personally; another vet may use something different. I use it when I want someone to give more than the labelled dose, for example, "Give Safeguard at 5X the labelled dose," when I want you to take the labelled dose and multiply it by 5. I'll also use it to tell you how many days I want you to give it. "Give Safeguard at 5X the labelled dose X 5 days."

Hopefully, these clarify some of the label directions that I, or another vet, may give you, and you can translate "Give Safeguard at 5X labelled dose PO SID X5 days," and "Give 4.5 cc/100lbs. (9mg/lb. dosing) LA200 IM EOD X 3 treatments."

Where Things Are Located on Your Animal

- **Rostrally** - towards the nose. Usually, I'm describing something on the face
- **Cranially** - towards the head
- **Caudally** - towards the back-end
- **Dorsally** - towards the back or spine area
- **Ventrally** - towards the belly
- **Laterally** - towards the side
- **Medially** - towards the center
- **Distally** - away from. Usually used to describe something on a limb
- **Proximally** - near to, another word used to describe something on a limb.

This is by no means an exhaustive list. But hopefully the next time you talk to your vet, you'll be able to understand better what we're talking about when we forget to translate.
Dr. H

What You Can Do to Make Your Veterinarian's Job Easier

My veterinary practice is purely mobile at this point. I travel to people's homes and farms to see their animals. This can go very smoothly, or it can be an exercise in frustration, depending on what is going on. Some people call me for an appointment and are prepared for my arrival. Others think they are prepared, but they really aren't. It's when owners aren't prepared

The Complete Granny Miller

that things can be very frustrating. Here are seven things you can do to help make my job or any traveling veterinarian's job easier.

Remember That I'm Coming
Yes, this has happened.
I arrive at a farm ready to work, only to find the owner isn't there. I call from the barn to make sure they are coming out …and they went to the store. When this happens, I still have to charge for my time. It's frustrating because I could have been helping someone else. If you have to be somewhere else, please just call and let me know!

Have the Animal Caught
Few things can waste my time more than waiting for an unprepared owner trying to catch the flighty animal. On any given day, I can have one farm call or several farm calls. When I'm ready to start, waiting for others to catch an animal is frustrating. This is especially true on days when I have more farms to visit, and it takes half an hour or longer to corral an animal. Have your animals in the barn or a pen. They don't have to be tied up. They just need to be accessible so we can reach them quickly.

Have the History Ready
Many times when I'm heading to a farm, I only get a quick summary over the phone; just enough to let me know what's happening. When I get to the farm, in order to help the animal fully, I need to know everything that has gone on, as well as that particular animal's history. It makes a big difference when I'm treating a downed animal to know how old it is, whether it's pregnant or not, if there was anything it may have eaten, or if there was an injury. It also helps to know how much time or money you are willing to put into the animal. Tell me ahead of time what the animal's purpose is. If it's a food animal, there are medications that I can't give them. I don't want to treat your animal with certain usable drugs if that animal is going to be slaughtered in X number of days.

Tell Me What You Have Already Done or Not Done
When I come to treat a sick animal, this is important for me to know. If you have already given a pain medication, please inform me, so that I don't give it again and overdose your animal. Tell me what you have already tried, so I can try something different or tweak what you have already done. And let me know which medications you have on hand, and which ones you don't. Speak up and tell me what you are comfortable doing for the animal and are able to do. That information can help me plan the treatment of your animal.

Have the Paperwork Ready
If you are taking your animals to shows, fairs, or selling them, it's helpful to let me know ahead of time where you are headed. With that information, I can double-check the requirements for you. If you are going out of state, there may be extra tests that the state you are traveling to requires. Also, there may be additional things a particular show wants you to do. Be

The Complete Granny Miller

prepared for that. Because if I find something that is required and you didn't know about it, I will let you know. Keep in mind that there are time frames for when I can do exams, tests, and CVIs before shows, fairs, or sales. Knowing ahead of time when to schedule me to come out is essential.

Tell the Truth

I can't tell you how unbelievably frustrating it is to go to a farm and have an owner flat-out lie to me about what is going on with the animal. I perform the physical exam and come up with a likely diagnosis, but the owner insists it can't be that because they did X, Y, or Z.

When you lie to the vet, we can't help you or your animal. Our job isn't to judge you. Our job is to help you give the best care to your animals. We can't help you take care of your animals if we don't know what's going on. I don't care that you made a mistake and forgot to give a medication or gave the wrong dose. I don't care if you forgot to lock the pen and they got out and into something they shouldn't have. I don't care that you should have been doing something, and didn't because of time or money constraints. I don't care about any of that.

What I care about is that I'm knowledgeable about everything, which allows me to treat your animal appropriately. With misinformation, I may treat your animal inappropriately, which terrifies me, because I could accidentally kill your animal. Just tell me the truth when I show up, and I will help you find the best way to take care of your animals.

Listen and Tell Me What You Don't Understand

Another frustration many vets have is clients who apparently listen to everything we tell them and then don't follow through on treatment. Most of us went through at least 8 years of school to become veterinarians. Many of us went to school even longer because of Masters Degrees and internships. We learned all the building blocks of diseases and treatment plans, so we can go out and help people care for their animals. When we have worked with you to come up with a treatment plan, and then learn on follow-up that you didn't complete the treatments needed to help your animal, that hurts us because it means you don't trust us.

When we are going over the treatment plan or explaining the reason why we are recommending this vaccine or this prevention management, if there is something you don't understand, just tell us. I, for one, am happy to go into more detail as to why I'm doing something. If you don't agree with something I suggest or recommend, tell me why, so we can work something else out.

Veterinarians work hard to help people take care of their animals. We didn't go into veterinary medicine for the money. We chose the profession because we care. We went through 8 or more years of school in order to practice in a profession we love. And we all swore an oath:

"Being admitted to the profession of veterinary medicine, I solemnly swear to use my scientific knowledge and skills for the benefit of society through the protection of animal health and welfare, the prevention and relief of

The Complete Granny Miller

animal suffering, the conservation of animal resources, the promotion of
public health, and the advancement of medical knowledge.
I will practice my profession conscientiously, with dignity and in keeping
with the principles of veterinary medical ethics.
I accept as a lifelong obligation the continual improvement of my
professional knowledge and competence."

So, when you call us to take care of your animals, remember that
veterinarians put a lot of time and effort into what we do. We will work with
you to help you care for your animals. But you have to help us out, too.
Respect our time, and we will respect yours. Please let us know what we
can expect to see at your place so we can plan accordingly. Let us know
what you have done already and what you are willing to do. Be honest with
us. Whenever I go onto a farm or to someone's home, I have sworn to do
my best. But I needyour help. Please keep in mind the things that you, as the
animal owner, can do to help me or any veterinarian do just that.
Dr. H.

Keeping Animals Warm in Cold Temperatures

Be it cold rain or a blizzard, farmers can't take a bad weather day off from
going out and taking care of their animals. When we're done with chores,
we can come back inside to the warmth of our homes. But what about the
animals who live outside? How do we keep them warm?

Well, for starters, most animals grow their own winter coats to help stay
warm.These coats tend to be fluffy, which traps body heat in. As long as an
animal is dry, these winter coats work very well. However, if animals get
wet, they get cold. And if the wind is blowing, it steals away body heat.
Providing shelter for animals to escape the wet and wind is essential.
Animal shelters, be it the barn, the shed, or the dog house, need to have an
area that blocks the wind.

But no shelter should be airtight. Shutting animals in can help keep them
warm, but without adequate airflow, it can lead to breathing troubles and
pneumonia. Shelters should block the wind and weather, while also
allowing some airflow to keep your animals' lungs healthy. Something else
to help keep animals warm in their shelters is deep, dry bedding.
Remember, damp and wet are the enemy. But deep bedding allows animals
to snuggle in out of the wind, creating a pocket of warmth.

Besides their winter coats and a good shelter, animals should also be eating
more feed during the winter to help them maintain a good body condition
and stay warm. Food provides energy to animals, which they use for
building muscle, laying down fat, growing, and other bodily functions. But
during the process of digestion, to get that energy, a lot of heat is produced!
Feeding your animals well also helps keep them warm. Good nutrition
keeps animals healthy and in good condition, which helps them withstand
cold temperatures. If they are too thin, they can't stay warm.

The Complete Granny Miller

Putting your hands on your animals to feel them and assess their body condition is essential during the winter. Those fluffy winter coats can actually hide a poor body condition. You should be able to feel ribs, but not get your fingers in between them. If you can get your fingers in between your animal'sribs, you need to increase their feed.

Having warm water available is also a good thing because it encourages animals to drink. Cold water chills them, so they drink less of it during the winter. Horses are especially vulnerable to colic during the winter when they don't take in enough water. We don't reach for ice-cold lemonade during the winter; we reach for hot tea and hot chocolate. Our animals are the same way.

So, the essentials for keeping your animals warm during the winter months are having good body conditions, good shelters that block the wind and wet, but not the airflow; deep bedding, and plenty of good feed with warm water if possible.

But when these things aren't enough, or you get an animal that needs a little more help staying warm, there are other things you can do.

Some animals can be blanketed to help supplement their winter coats. Horse blankets come in all different sizes, and foal blankets can fit goats, llamas, and alpacas. Just make sure the coat stays dry, because again, wet and damp are the enemy during winter. Heat lamps can also be a help, but you have to be careful. Keep the light and the cords up where the animals can't reach them. Broken lights can lead to fires and chewing on cords to electrocution. Lamps should be positioned in a corner or other sheltered area out of the wind. Animals should be able to get under the heat lamp if they want to, but also need to be able to get away from it if they get too warm.

Keeping animals warm in the winter can be a challenge, but keeping these things in mind can help you keep them happy and healthy.

Dr. H

The Complete Granny Miller

Over the years, I have received more than a few emails thanking me for the following information on livestock and small animal euthanasia. It seems there's not a lot of information regarding this topic on the Internet. When this information is necessary, it's usually needed fast to prevent further animal suffering.

Small Animal and Livestock Euthanasia

Tibby is a young barn cat that will need to be euthanized within the next few days. She is less than a year old and is suffering from an aggressive form of cancer. For the past week, I've been spending extra time with her and feeding her all the milk and cheap bologna she cares to eat. She doesn't seem to be in any pain just yet. I'm watching her carefully for the first signs of pain, or a change in her behavior, or for the tumor to begin to rupture. At the first hint of a change, my husband or I will euthanize her quietly here on the farm.

Without a doubt, one of the most unpleasant but vital homestead skills is the ability to quickly and painlessly euthanize sick or suffering animals and livestock. For most animals, the preferred method on this farm is a well-placed bullet to the front or back of the head while the animal is eating or distracted in some way. That's how Tibby will be released. She will be shot from behind while she is eating. Her death will be instantaneous, and she will never know any pain.

On our farm, we use small-caliber bullets for small animals and larger-caliber bullets for livestock. Chickens, ducks, and other poultry are not shot. But instead, they are quickly euthanized with a broomstick. We never use the services of a veterinarian for euthanasia due to cost and time considerations, and because it is less stressful for an animal to be put down by someone they know and trust.

I prefer a .22 caliber bullet for cats, small dogs, goats, and light pigs. I use a .38 caliber hollow point bullet for sheep, cattle, horses, and heavy hogs. My husband prefers a .45 caliber LC for larger animals.

It doesn't matter if the shot is made from a rifle or a handgun. However, a rifle produces a bullet with a higher velocity. That can be a critical consideration in some situations.

I almost always use a handgun when I have to destroy an animal because it's what I'm comfortable with. But at times, it can be safer for the shooter to use a rifle if the animal is large and in pain. An animal in pain is unpredictable and can be quite dangerous. Whenever possible, I restrain and remove the animal from the other animals so that they don't witness the

The Complete Granny Miller

killing. Some people say it doesn't matter. But I think that it does. Animals understand a lot more than we sometimes give them credit for.

When euthanizing an animal, the most important thing to keep in mind is safety for the shooter and for any other creatures nearby. It is safest to have bystanders stand behind the shooter and well back away from the animal. If at all possible, try to move the animal outdoors and avoid taking the shot in the barn if it can be helped. A ricochet bullet is unlikely. But take care that nothing obviously hard or solid is in the way or the line of fire. That said, if it is too stressful or upsetting to the animal to be moved, I will shoot it in the barn.

When outdoors, I try to take the shot while standing behind the animal, and facing downhill if I'm on a hill. That's because an animal will often lunge forward when first shot, and it is easier for the shooter to back up. Almost always, animals will jerk, thrash, and twitch uncontrollably when shot in the head. It is imperative to be able to step out of the way quickly so as not to be accidentally hurt.

The most effective headshot is a shot that is taken 3 inches to 12 inches away from the back or the front of the head. **Never with the muzzle of the gun placed directly on the head.** A little extra distance allows the shooter to shift if the animal moves. The shot should be aimed directly downward, between the ears, when standing behind the animal. When standing in front of the animal, the shot is placed between the eyes or mid-line on the forehead. The angle of the shot and placement depend on the species and the shooter's position. This is where it's essential to be aware of the fundamental physiological differences in livestock and small animals. Skull shape is not the same in all animals. Take the time to learn ahead of time how the animals you keep and are responsible for are put together.

The more precisely a bullet is placed into the center of the brain, the more catastrophic the tissue damage. Catastrophic damage results in a merciful and quicker kill. It's a case of lights on - then lights off. There is no pain for the animal. It's a complete short circuit from the brain to the body. If you are unsure about the exact bullet placement, a larger caliber bullet can reduce the margin of error. Two shots fired into the skull in rapid succession will kill or fatally stun most large farm animals.

With chickens, ducks, and other poultry, I believe the most merciful and quickest kill is by way of cervical dislocation with a broomstick. I first restrain the chicken and hold its wings close to its body. Next, I place the chicken or duck, with its beak/bill and breast, side down on a hard surface, such as a cement pad or sidewalk. The broomstick is positioned so that it spans directly across the back of the chicken or duck's neck, where the head meets the neck. I then step quickly on the left side of the broomstick, followed by the right side. Next, I firmly and quickly pull the chicken's body by its feet towards me, away from the head and broomstick. By stepping on the broomstick while it spans the bird's neck and then pulling or jerking the body backwards, the spinal cord is severed from the brain, and death is instantaneous.

The proper disposal of euthanized animals is an important consideration. On this farm, all animals are either buried or taken to the woods and left

The Complete Granny Miller

exposed so that other animals can make good use of them. If you know ahead of time that you will need to euthanize an animal, it is helpful and practical to have the grave dug in advance or have a plan for the removal of the body. If you're going to bury an animal, it's important to be sure to bury it deep enough. Graves should be at least 3 feet deep for most small animals. For large livestock, graves should be at least 5 feet deep and wide enough. A backhoe and a set of chains are real time savers for large farm animals. Remember to keep the graves away from wells and other water sources.

No conversation about animal euthanasia would be complete without a mention of the human emotions that are involved. Speaking from personal experience, I have found that there's a profound sense of regret, sadness, and emptiness when any animal has to be destroyed. A feeling of interior hollowness and the stillness, as well as the absolute finality of death, are always present. Often, there is self-blame, whether or not it is merited. When the animal is a pet or there is a strong emotional attachment, euthanasia can be exceptionally hard. It's at that time that personal courage, bravery, and faith are necessary. Because euthanizing a pet can be difficult, many people will elect to use the services of a veterinarian or call a trusted friend. There's no shame in asking someone else to shoot your dog or horse. We unfortunately live in a society that denies death and anthropomorphizes animals. So, there are bound to be problems when we're faced with the euthanasia of our pets and animals that we love. Often, emotions will cloud good judgment. Sadly, many animals have been held onto way past the time when they should have been released and allowed to pass away.

But sometimes euthanasia can be an easy choice with few regrets. I have found this to be especially true with large livestock. When an animal is obviously suffering, and there is no possible hope or remedy for the situation, it is easy to take the shot. At those times, courage is not needed. Only mercy is required.

Mercy is a gift that we, as humans, can bestow upon the animals that serve and depend on us. Mercy helps me to find my target and to remain calm, detached, and determined while I do what I must. I always say a prayer right before I take the life of any animal. I pray that God will steady my hand and give the animal a quick and painless death. I also pray for forgiveness. Never once have I killed an animal without being cognizant that death is the cost for this earthly life. And that one day I too will be required to pay the price.

The Complete Granny Miller

Chapter 7

HOMESTEAD WISDOM & FOLKLORE

"God is best known in not knowing him." - Saint Augustine.

Folk Wisdom in the Modern World

As a garden farmer, it seems to me that the modern Western world shares a collective illusion. The illusion is the acceptance and conviction that observable and predictable natural phenomena known to gardeners and farmers for thousands of years are without merit simply because they are not scientifically verified or reproducible. However, modern science, conceived and organized solely on deductive and inductive reasoning, cannot explain away centuries of intuitive insight and understanding regarding the natural world. For hundreds of generations, the survival of our ancestors depended upon traditional agrarian folk knowledge, experience, and wisdom. This wisdom was transmitted orally and sometimes by images to each successive generation. Without such pragmatic knowledge and insight, we would not be here today. Our forefathers and foremothers would have starved to death or perished from exposure to the elements.

In a world filled with grocery stores, ready-to-wear clothing, and TV weather reports, it's easy to overlook the vulnerabilities, risks, and perils of human life. What follows is a collection of homestead wisdom, old-time proverbs, and just plain wacky farm life superstitions. This information was freely transmitted to me by the two generations of garden farmers and homesteaders that came before me. I'm now discharging my obligation and passing it on to you.
Guard it carefully. Because once lost, it's gone for good.

50 Old Time Weather Proverbs & Signs

After years of experience, I can usually predict rain just by walking on the morning grass, watching the behavior of my cats and cows, or just by opening a dresser drawer. What follows is a list of my favorite weather folk sayings. And like many folk proverbs, you'll find more times than not they have real merit and value. In fact, for the most part, I've found the weather wisdom

The Complete Granny Miller

that follows to be more accurate than a meteorologist using
computer models or satellite imagery.

- When a hornet's nest is built high in
  the top of a tree, it indicates a mild
  winter is ahead. Nests built close to
  the ground indicate that a harsh winter
  is coming.
- The higher the clouds, the better the weather.
- If a cat washes her face over her ear, the
  weather is sure to be fine and clear.
- Clear moon, frost soon.
- When leaves fall in early
  autumn, the winter will be mild.
  When leaves fall later, winter
  will be severe.
- March comes in like a lion, and goes out like a
  lamb.
- When ants travel in a straight line,
  expect rain; when they are scattered,
  expect fair weather.
- If the first snow falls on unfrozen ground,
  expect a mild winter.
- If bees stay at home, rain will
  soon come; if they fly away, fine
  will be the day.
- A year of snow, a year of plenty.
- Dust rising in dry weather is a sign of
  approaching change.
- Rainbow at noon, more rain soon.
- Flowers blooming in late autumn are a sign of a
  bad winter.
- If cows lie down and refuse to go to
  pasture, you can expect a storm to
  blow up soon.
- The darker the woolly caterpillar's
  coat, the more severe the winter will
  be. If there is a dark stripe at the head
  and one at the end, winter will be
  severe at the beginning, become mild,
  and then worsen just before spring.
- When the grass is dry in the morning light, look
  for rain before nightfall.
- If sheep ascend hills and scatter, expect clear
  weather.
- A warm November is the sign of a bad winter.
- When the chairs squeak, it's of rain they speak.

The Complete Granny Miller

- When clouds appear like rocks and towers, the earth will be washed by frequent showers.
- If birds fly low, then rain we shall know.
- Evening red and morning grey are two sure signs of one fine day.
- The first and last frosts are the worst.
- The winds of the daytime wrestle and fight, longer and stronger than those of the night.
- When down the chimney falls the soot, mud will soon be underfoot.
- Rain before seven, fine before eleven.
- No weather is ill if the wind be still.
- Cold is the night when the stars shine bright.
- When a rooster crows at night, there will be rain by morning.
- Dandelion blossoms close before a rain.
- When clouds look like black smoke, a wise man will put on his cloak.
- A cow with its tail to the west makes the weather best; a cow with its tail to the east makes the weather least.
- The moon and the weather may change together, but a change of the moon will not change the weather.
- The sudden storm lasts not three hours.
- Chimney smoke descends, our nice weather ends.
- A rainbow in the morning is the shepherd's warning. A rainbow at night is the shepherd's delight.
- Three days of rain will empty any sky.
- When smoke hovers close to the ground, there will be a weather change.
- A ring around the sun or moon means rain or snow coming soon.
- Bees will not swarm before a storm.
- The more cloud types present, the greater the chance of rain or snow.
- Catchy drawer and sticky door, coming rain will pour and pour.

The Complete Granny Miller

- When the wind blows from the west, fish bite best. When it blows from the east, fish bite least.
- When leaves show their underside, be very sure that rain betides.
- Birds on a telephone wire predict the coming of rain.
- When the ditch and pond offend the nose, then look out for rain and stormy blows.
- Pigs gather leaves and straw before a storm.
- Trout jump high when rain is nigh.
- Red sky at morning, sailor take warning; red sky at night, a sailor's delight.
- When the night goes to bed with a fever, it will awake with a wet head.

Tiny Chicken Eggs – A Natural Phenomenon with a Spooky History

When I went to collect eggs yesterday, I found a dwarf egg sitting in the nest boxes along with the regular-sized eggs. I thought to throw the egg over the house, but instead decided to tempt Fate, and I brought it indoors. Tiny or miniature-sized eggs in standard-sized hens are the natural result when a small bit of reproductive tissue or other small foreign mass enters the hen's oviduct and triggers the regular formation of an egg. Inside the hen's body, the bit of tissue or foreign mass is treated exactly like a normal yolk. It is swathed and enveloped in albumen, membranes, and a shell, and is eventually passed from the hen's body. When it is laid, it looks just like a regular chicken egg except that it is tiny.

These types of malformed eggs have been known for centuries as a 'cock egg'. Most often, these little eggs contain only the white of the egg and no yolk. Usually, the shells are harder to break than those of a normal egg. Cock egg is a synonymous term for any type of abnormal egg.

In folk tradition, a cock egg was understood to have been laid by a rooster or cock and not a hen, and it was a cause for grave concern. Cock eggs, according to different folklore beliefs, bring bad luck or illness if they are brought into the house. That's because a cock egg is believed to have malefic and magical powers. They are reputed to be of value to sorcerers and magicians for mixing magical potions and casting spells. The way the story goes is that if a toad, serpent, or witch, at the behest of Satan, incubates a cock egg, the resulting hatchling will be a cockatrice or a basilisk. A cockatrice or basilisk is an ancient winged monster with a serpent's body and a rooster's head that can kill and destroy by its breath and glance.

The Complete Granny Miller

During the Middle Ages, it was self-evident to most intelligent people that a cock egg was the work of the devil. Animals as well as people could be in league with Satan. And in 1474, a chicken passing for a rooster in Basle, Switzerland, was put on trial and condemned to be burned at the stake for "the heinous and unnatural crime of laying an egg".

American author and educator, E.V. Walter, in his essay, *Nature On Trial - The Case of A Rooster That Laid An Egg,* writes, " the execution took place with as great a solemnity as would have been observed in consigning a heretic to the flames, and was witnessed by an immense crowd of townsmen and peasants."

A cock egg has also been called a 'witch egg' since the Middle Ages, and was called a 'fairy egg' during the mid and late Victorian era. In Scotland and elsewhere in Europe, a cock egg is sometimes also called a 'wind egg'. In recent times, here in the U.S., these types of deformed eggs are sometimes called 'fart eggs'. I suppose language really does reflect cultural ideals and concerns.

Superstition instructs that the best way to protect against the evil of a cock egg is to throw the malformed egg over the roof of the house and smash it on the other side, which, of course, I didn't do. So now I guess we'll just have to wait and see what happens next. But I'm not too worried.

"A Year of Snow, a Year of Plenty"

It's a good old weather proverb and a truism founded upon practical agrarian experience that's been passed down through many generations of farmers and gardeners. I thought you might be interested in knowing a few of the reasons behind this old-time weather adage.

Lots of Snow Means Lots More Water

One of the reasons that a bountiful fall harvest follows a winter with heavy snowfall is that there's plenty of fresh, clean groundwater for food production. After a winter of substantial snow accumulation, underground aquifers and household wells are recharged during the spring melt and thaw. Unconfined fresh water aquifers are vital to all aspects of agriculture and rural life. An abundance of fresh water will help to ensure that field crops, orchards, and home vegetable gardens will have plenty of moisture and water for the coming growing season.

Plants Are Better Protected

Deep snow that lies on the ground throughout the winter is of benefit to fieldcrops, orchards, small fruits, and perennial garden plants. Heavy snow acts like a frozen mulch and insulates the soil and plants. During the winter months, plants are protected from the destructive cycle of 'freeze, thaw, and heave' by a continuous cover of snow. It's not the bitter cold that will damage plants as much as it is the wind and ground heaving. The repeated thawing and then refreezing of the ground tends to heave plants up from the ground, creating air pockets that leave roots exposed. This makes plants

The Complete Granny Miller

vulnerable to winter kill, wind damage, and drying out.

Fruit Trees Are Slower to Bloom
A long winter with heavy snow and sustained cold temperatures will tend to protect orchards and small fruits from blooming too early. Lots of snow and cold weather will slow down the flowering of fruit trees and allow them to bloom when it's safe. Often, a brief warm spell in early spring will cause fruit trees to bloom and flower too early. If an unexpected killing frost catches fruit trees in bloom, fruit production can be severely limited or destroyed for that year.

Destructive Pests Are Controlled
Cold weather and heavy snow keep insect pests and internal parasite populations under control. Fewer insect pests mean less pest damage to fields, orchards, and gardens, and more food for people. When internal parasites aren't reproducing as quickly due to cold weather, it makes for better feed conversion, and healthier and happier livestock and farmers.

Gardening with the Moon

The term 'planting by the signs' is a colloquial or folk expression for the ancient practice of scheduling farming and gardening work based on the moon's position in the zodiac. The term 'planting by the signs' is a colloquial expression or folk term for the ancient practice of timing agricultural and gardening tasks by the moon's astrological position in the zodiac.

For our ancestors, the understanding and application of natural lunar cycles and rhythms to their lives was literally a matter of life and death. By applying the primeval principles passed down to them by each preceding generation, our forefathers and foremothers managed to survive famine and disease. It wasn't important for our agrarian ancestors to understand the science of why something worked. What mattered to them was that it did work. Predictable lunar cycles and phases were an essential part of life for our forebears. It's the primary reason almanacs were so widely used. It's also the reason why, next to the Bible, almanacs were the second most important book in nearly all Christian households. For those of you who may believe that planting by the signs is pure superstition or backwards nonsense, I would encourage you to suspend judgment and experiment for yourself. Engage in a closer scrutiny of the natural world. You may just be surprised at what you discover.

Up until the modern scientific era, most people in the world understood the importance of lunar cycles. Countless generations have used the information that follows for their benefit and survival. Agricultural or natural astrology is a highly involved subject. I can't do the topic justice in this short section. Instead, the purpose of this explanation is to provide a brief summary of the concepts and ideas behind moon sign planting, and to offer a general overview of its application in modern family gardens.

The Moon

"Let there be lights in the firmament of the heaven to divide the day from the night; and let them be for signs, and for seasons, and for days and for years."
-Genesis 1:14

Every month, the moon goes through a four-stage lunar cycle. It takes approximately 29.5 days for the moon to complete its full cycle. During the monthly lunar progression, the moon's phases are divided into four parts or quarters.

NEW MOON → 1st Quarter → 2nd Quarter → FULL MOON →3rd Quarter→ 4th Quarter →NEW MOON

Each phase or quarter lasts slightly longer than a week.The moon's quarters divide the two significant visual manifestations of the moon, known as the **New Moon** and the **Full Moon**.

- The New Moon is the interval of the lunar cycle when the moon is not visible in the night sky.
- The Full Moon is the period when the moon is completely round and illuminated.

The monthly lunar cycle begins with the **New Moon**. This 7-day part of the lunar cycle begins in complete darkness. During the actual day of the New Moon, the moon is not visible in the night sky.

The first day after the New Moon starts the part of the lunar cycle known as the **1st Quarter**. During the 1st Quarter, the moon gradually begins to increase its visibility and light. If you look up at the moon during this time, you will notice that the moon is becoming increasing crescent-shaped each night.

The crescent-shaped moon in the night sky is formed by a shadow from the light of the sun's light. The tips or points of the crescent moon always point to the sun. The top and bottom tips of the moon crescent are called the 'horns' of the moon. In the Northern Hemisphere, the horns of the moon point to the left as the light of the moon increases. In the Southern Hemisphere, the crescent direction is reversed and the horns point to the right. The horns of the moon are a significant concept in folklore and myth. During the 1st and 2nd Quarters, the increase in the light of the moon is known as a **'waxing moon'**.

The Complete Granny Miller

As the moon begins to increase in size and outgrow its crescent shape, it moves into the 2nd Quarter of the lunar cycle. During the 2nd Quarter, the moon is continually increasing in light. Soon it will become the complete round orb known as a **Full Moon**. During the Full Moon, the moon reaches its zenith and is at its greatest maximum light. The Full Moon marks the halfway point of the monthly lunar cycle.

Beginning the day after the Full Moon, the **3rd Quarter** of the lunar cycle commences. At this time, the moon gradually begins to decrease in light. During this period of the lunar cycle, the horns of the moon now point to the right if you are located in the Northern Hemisphere, and they point to the left if you are in the Southern Hemisphere.

The **4th Quarter** is the period of the lunar cycle when the moon returns to darkness and to the New Moon phase, which will begin the cycle once more.

The Zodiac

In order to understand old almanacs or to benefit from pre-World War II gardening and farming lore, it is necessary to have a rudimentary foundation and understanding of natural astrology. Natural astrology has little to do with sun sign horoscope nonsense found in newspapers and magazines, or on esoteric websites for the confused and lovelorn. Instead, natural astrology is the ancient branch of astrology that focuses on the natural world. Nature, not humans, is the concern here.

The weather, tides, volcanic eruptions, plagues, earthquakes, animal behavior, and predictable cycles of agriculture are the focus of this branch of astrology. Claudius Ptolemy (AD 100–170) was a Greco-Egyptian writer, mathematician, astronomer, geographer, and astrologer who authored a seminal astrological work called the Tetrabiblos. The Tetrabiblos is a compilation of four related books on astrology by Ptolemy, drawn from earlier ancient works. Two of those books helped to form the foundation of all Western natural astrology. Ptolemaic astrology was taught in European universities well into the 17th and 18th centuries.

Believe it or not, ancient astrology is considered to be the mother of all sciences. It is the forerunner of modern astronomy and medicine. Less than 200 years ago, every farmer, medical doctor, and rational, well-educated person had more than a passing knowledge of the basic principles of natural astrology.

For thousands of years, ancient astronomers observed that many of the constellations they studied were evenly spaced in an imaginary band along the sun's apparent yearly path in the sky. Astronomers divided this circular belt evenly into twelve sections, with each section measuring 30°. This ever-changing seasonal and imagined ring of stars above our heads in the night sky is called the **zodiac**. Each of the twelve 30° sections or **signs** of the zodiac is represented by a living creature, except for one - Libra. The seventh sign, Libra, is an inanimate object and is symbolized by a balanced scale. The name zodiac is derived from the Greek and then Latin word for 'circle of animals'.

The Complete Granny Miller

The 12 Signs of the Western Zodiac

- Aries - The Ram
- Taurus - The Bull
- Gemini - The Twins
- Cancer - The Crab
- Leo - The Lion
- Virgo - The Virgin
- Libra – The Scales
- Scorpio - The Scorpion
- Sagittarius - The Archer
- Capricorn - The Goat
- Aquarius - The Water Bearer
- Pisces - The Fishes

The zodiac as a circle begins in Aries and ends in Pisces. Every year, the sun passes through all twelve signs of the zodiac and spends about 30 days in each sign. Every month, the moon, just like the sun, also passes through all twelve signs of the zodiac. The moon spends just under 2 ½ days in each sign.

Qualities of the Signs

Each of the twelve zodiac signs is associated with a different characteristic, quality, or trait. To understand the quality or attribute of the twelve signs of the zodiac, they are sorted into four groups. Each group contains three signs that correspond with four of the five ancient elements:

- Water
- Fire
- Earth
- Air

An awareness and consideration of the qualities and attributes of the zodiac signs are fundamental for proper insight and the application of natural astrology. When it comes to gardening, food preservation, and animal husbandry, the moon phase is more important than the zodiac sign. Nevertheless, it's essential to recognize why certain days of the lunar month are traditionally considered to be favorable or unfavorable for particular tasks. Without a basic understanding, it's pretty much impossible to read and comprehend an ordinary garden almanac.

The Water Signs

Water signs are said to be feminine, wet, productive, nutritive, or fruitful. The water signs are Cancer, Scorpio, and Pisces.

The Fire Signs

Fire signs are said to be masculine, barren, unfruitful, harsh, and dry. The firesigns are Aries, Leo, and Sagittarius.

The Complete Granny Miller

The Earth Signs
Earth signs are said to be earthy, sturdy, stable, substantial, and feminine. The earth signs are Capricorn, Taurus, and Virgo.

The Air Signs
Air signs are said to be masculine, fickle, fluctuating, vacillating, and airy. The air signs are Libra, Aquarius, and Gemini.

Attributes of the Signs
The zodiac is further categorized by the attributes of the signs. The twelve signs are divided into three categories, with each category containing four signs. The three categories of the zodiac are known as Cardinal, Fixed, and Mutable.

Cardinal Signs
The Cardinal signs are those signs that generate, produce, or begin a new season or condition. Hence, Aries, Cancer, Libra, and Capricorn are the signs that herald each new season. Aries begins spring. Cancer begins summer. Libra begins autumn. Capricorn begins winter. The Cardinal signs correspond to the four points of the compass: north, east, south, and west.

Fixed Signs
The Fixed signs have a settled, stubborn, or unmoving quality to them. Taurus, Leo, Scorpio, and Aquarius are fixed signs and are ascribed to the established seasons of spring, summer, autumn, and winter. Fixed signs are stable and committed. They are sometimes considered to be 'earthquake signs' due to the belief that earthquakes occur shortly before or after a lunar eclipse when the moon is in a fixed sign.

Mutable Signs
The Mutable signs are unstable and unpredictable. They are symbolic of the inevitability of change. The zodiac signs of Gemini, Virgo, Sagittarius, and Pisces are the signs of the zodiac that are assigned to the months before one season ends and another season begins. Mutable signs are transformative. Winter breaks with Pisces. Summer comes after Gemini. Virgo signals autumn. Winter begins after Sagittarius. The mutable signs are considered to be the reuniting, resolving, and reconciling energies of the universe and cosmos.

The Complete Granny Miller

When & What to Plant

"To everything there is a season, and a time to every purpose under heaven. A time to be born and a time to die. A time to plant, and a time to pluck up that which is planted." Ecclesiastes 3:1

Knowing the correct sign to plant in is essential. But knowing what phase of the moon to plant in is even more critical. By combining good gardening practices with the fundamental principles of the lunar cycle, along with the moon's zodiac position, modern gardeners and farmers can use the wisdom of our ancestors to grow better food and more beautiful flowers.

In general terms, things tend to increase in the light of the moon and then decrease in the dark of the moon.
Plants or vegetables that grow above ground should be planted during the 1st and 2nd quarters of the moon when the moon's light is increasing. Broccoli, asparagus, cabbage, peppers, tomatoes, squash, cucumbers, green beans, and corn are all planted during the 1st and 2nd quarters of the moon.

Plants, grains, herbs, or flowers that carry their seeds outside of the fruit or vegetable do best when planted in the 1st quarter. Corn, broccoli, wheat, marigolds, dill weed, and chamomile are examples of plants that produce exterior seeds.

Plants, flowers, grains, or herbs that have interior seeds are most often planted in the moon's 2nd quarter. Squash, watermelon, green peppers, tomatoes, green beans, poppies, pumpkins, and peas are examples of plants that produce interior seeds. They should be planted during the moon's 2nd quarter.

Cucumbers are the exception to the rule. They should be planted during the moon's 1st quarter.

During the 3rd and 4th quarters, the light of the moon is decreasing. Crops that are grown underground for their roots or tubers should be planted at that time. Potatoes, turnips, carrots, and beets are best planted during a waning moon up until a few days before the New Moon.

Bulbous flowers like tulips, crocus, daffodils, and hyacinths are planted during the moon's 3rd quarter. The best results are seen when underground crops or bulbous plants are planted in a fruitful sign and often in an earth sign.

Be aware that the closer a planting is done to the New Moon, the crop results are somewhat less favorable. No crops should be planted on the day of the New Moon. In general, gardening activities to destroy garden weeds and pests are more effective when done during the moon's 3rd and 4th quarters. Such activities, when combined with a dry or barren sign, result in weeds, pests, and parasites, which are destroyed much more easily.

The Complete Granny Miller

Aries

Aries is the first sign of the Western zodiac. It is the sign of the Ram and rules the head. It is a masculine, barren, dry fire sign. It is the Cardinal Fire sign and is associated with the planet Mars and the god of war. This is a good sign to destroy weeds and pests. It is a reliable sign to harvest fruit and root crops during the moon's 3rd & 4th quarters.

Taurus

The second sign of the Western zodiac. It is the sign of the Bull and rules the neck or throat. It is a moist, earthy, and feminine sign. Taurus is the fixed Earth sign and is associated with the planet Venus and the goddess of Beauty. Taurus is used to plant crops and flowers when hardiness or firmness is a principal factor. When the moon is in Taurus during the 1st and 2nd quarters, cabbage, lettuce, kale, and other leafy vegetables are planted. Hay, beans, peach, and pear trees can be planted in Taurus. During the 3rd and 4th quarters, root vegetables like potatoes, radishes, beets, and carrots are planted. Taurus is useful for planting all bulbous plants and flowers. Daffodils and tulips can be planted in Taurus.

Gemini

The third sign of the Western zodiac. It is the sign of the Twins and rules the arms and chest. Gemini is the Mutable Air sign and is associated with the planet Mercury and the messenger god. Gemini is masculine, dry, and barren. This is a good sign for cultivating and stirring up the soil to destroy weeds. Lawns can be mowed at this time to retard growth, and root crops and fruit can be harvested in this sign during the moon's 3rd and 4th quarters. Gemini is never a good time for transplanting or planting.

Cancer

The fourth sign of the Western zodiac. It is the sign of the Crab and rules the breast and stomach. Cancer is the Cardinal Water sign and is associated with the moon, and with women, motherhood, and nurturing. Cancer is fruitful, moist, and feminine. Of all the signs, Cancer is the most productive and is best used for most plantings. It is an advantageous sign for watering the garden or for irrigation. When the moon is in Cancer during the 1st and 2nd quarters, it is an excellent time to graft fruit trees or begin potted stem cuttings. All root vegetables can be planted at this time if the moon is in the 3rd or 4th quarter. Alfalfa, melons, wheat, and roses all do well when planted at this time. Cancer is such a moist and watery sign that it is not a good time to harvest potatoes, dry herbs, tomatoes, or onions. Things tend not to dry or cure properly in Cancer.

Leo

The fifth sign of the Western zodiac. It is the sign of the Lion and rules the heart and the back. Leo is the Fixed Fire sign and is associated with the sun and with the concept of father. It is a masculine sign that is deemed to be barren, dry, and fiery. In fact, of all the signs, Leo is judged as the most barren. Like Aries, it is considered to be a 'killing sign'. Leo is principally

The Complete Granny Miller

used to kill weeds, destroy roots, and girdle trees. Leo is used to mow lawns to retard growth, and is used to harvest fruit and root crops. Herbs, garlic, and onions dry the fastest when harvested in Leo.

Virgo
The sixth sign of the Western zodiac. It is the sign of the Virgin and rules the bowels and intestines. Virgo is the Mutable Earth sign, associated with the planet Mercury and the transition into adolescence. It is a feminine and moist sign, but is considered barren. Virgo is not used for planting. Instead, like Gemini, Virgo is helpful for cultivation. That said, many old-timers believe that Virgo is a good sign to plant vines and for planting certain flowers. Honeysuckle, moon vine, morning glory, peonies, gladiolas, and iris are sometimes planted in Virgo.

Libra
The seventh sign of the Western zodiac. It is the sign of the Balanced Scale and rules the kidneys and bladder. Libra is the Cardinal Air sign and is associated with the planet Venus and beauty. It is a masculine sign that is moist and semi-fruitful. Interestingly, Libra is the only sign in the zodiac that is not a living creature, but instead is an inanimate object. Libra is used for planting crops where good fleshy growth and roots are desirable. Barley, corn, wheat, rice, oats, millet, spelt, and rye can all be seeded and planted in the 1st and 2nd quarters of Libra. Grain planted at this time will produce a judicious and reasonable harvest. Libra is the best sign for planting all vines and is especially useful for fragrant and beautiful flowers. Old-timers often refer to Libra as "the sign of the reins".

Scorpio
The eighth sign of the Western zodiac. It is the sign of the Scorpion and rules the reproductive system and genitals. Scorpio is a feminine sign, which is fruitful, moist, and watery. It is the Fixed Water sign that is associated with the planet Mars, and in modern times with the planet Pluto. Next to Cancer, Scorpio is counted to be almost as fertile. When a Cancer day cannot be used, a Scorpio day can work just as well. Scorpio days during the 1st and 2nd quarters are ideal for planting beans, peas, eggplant, cucumbers, asparagus, parsnips, carrots, pumpkins, gourds, and all types of melons. As with Cancer, do not harvest potatoes or other root crops during Scorpio days because they will not keep in storage.

Sagittarius
The ninth sign of the Western zodiac. Sagittarius is the sign of the Archer and rules the hips, thighs, and buttocks. Sagittarius is a dry, barren, and masculine sign. It is the Mutable Fire sign and is associated with the planet Jupiter. Sagittarius is used like other fire signs for destroying weeds and pests. It is a reliable sign for cultivation and can be a good sign for planting garlic, onions, and chives. New hay fields are often planted in Sagittarius with a good result.

The Complete Granny Miller

Capricorn

The tenth sign of the Western zodiac. Capricorn is the sign of the Goat and rules the knees, bones, hair, teeth, and skin. Capricorn is the Fixed Earth sign and is associated with the planet Saturn and time. It is a practical and earthy feminine sign that is a little drier than Taurus. During the 3rd and 4th quarters of the moon, Capricorn is useful for planting potatoes, root crops, and bulbous plants and flowers. Evergreen trees, shrubs, hedges, and living fences can be planted in Capricorn during the 1st and 2nd quarters of the moon. Plants that are used for their seeds, like buckwheat and sunflowers, can be planted in Capricorn.

Aquarius

The eleventh sign of the Western zodiac. Aquarius is the sign of the Water Bearer and rules the legs, calves, and ankles. In modern times, Aquarius is associated with the planet Uranus. It is the Fixed Air sign that is dry, barren, airy, and masculine. Aquarius is good for destroying weeds. During the 3rd and 4th quarters of the moon, Aquarius can be good for harvesting fruits, some medicinal herbs, and root crops.

Pisces

The twelfth sign of the Western zodiac. Pisces is the sign of the Fishes and rules the feet and toes. In modern times Pisces is associated with the planet Neptune. Pisces is the Mutable Water sign and is moist, feminine, and fruitful. Like Cancer and Scorpio, Pisces is one of the best signs to plant in. During the 1st and 2nd quarters of the moon, all fruit and deciduous trees can be planted when the moon is in Pisces. Pisces is particularly good for all root growth, for flowers, and for seeding new lawns.

Moon Planting Basics

- Plants that grow above the ground should be planted in the increasing light of the moon (waxing)
- Plants that grow beneath the ground should be planted during the decreasing light of the moon (waning)
- Plants that carry their seeds on the outer part of the plant should be planted during the moon's first quarter.
- Plants that carry their seed inside the fruit or vegetable should be planted during the moon's second quarter

In general, plant and transplant in moist, feminine Water and Earth signs. Destroy weeds and pests in Fire and Air signs.

The Complete Granny Miller

ZODIAC MOON SIGN & MOON PHASE FOR PLANTING COMMON VEGETABLES

CROP	BEST SIGN TO PLANT IN	MOON QUARTER TO PLANT IN
Asparagus	Cancer, Scorpio, Pisces, Taurus	First & Second Q
Barley	Cancer, Scorpio, Pisces, Libra	First & Second Q
Beans	Cancer, Scorpio, Pisces, Taurus (*Gemini)	Second Q
Beets	Cancer, Scorpio, Pisces, Taurus	Third Q
Broccoli	Cancer, Scorpio, Pisces, Taurus	First & Second Q
Brussels Sprouts	Cancer, Scorpio, Pisces, Taurus	First & Second Q
Buckwheat	Cancer, Scorpio, Pisces, Capricorn	First & Second Q
Cabbage	Cancer, Scorpio, Pisces, Taurus	First & Second Q
Cantaloupe	Cancer, Scorpio, Pisces, Taurus, Libra	First & Second Q
Carrots	Cancer, Scorpio, Pisces, Taurus	Third Q
Cauliflower	Cancer, Scorpio, Pisces, Taurus, Libra	First Q
Corn	Cancer, Scorpio, Pisces, Libra	First & Second Q
Cucumbers	Cancer, Scorpio, Pisces	First Q
Eggplant	Cancer, Scorpio, Pisces, Libra	First & Second Q
Garlic	Cancer, Scorpio, Sagittarius, Pisces	Second & Third Q
Gourds	Cancer, Scorpio, Pisces, Libra	First & Second Q
Hay	Cancer, Scorpio, Pisces, Taurus	First & Second Q
Herbs	Cancer, Scorpio, Pisces	First & Second Q
Horseradish	Cancer, Scorpio, Pisces, Taurus	Second & Third Q
Kale	Cancer, Scorpio, Pisces, Taurus	First & Second Q
Lettuce	Cancer, Scorpio, Pisces, Taurus	First & Second Q
Millet	Cancer, Scorpio, Pisces, Libra	First & Second Q
Oats	Cancer, Scorpio, Pisces, Libra	First & Second Q
Onion Sets	Cancer, Scorpio, Pisces, Taurus, Sagittarius	Second & Third Q
Parsnips	Cancer, Scorpio, Pisces, Taurus	Second & Third Q
Peas	Cancer, Scorpio, Pisces, Taurus, Libra	Second Q
Peppers (Hot)	Scorpio, Sagittarius	Second Q
Peppers (Sweet)	Cancer, Scorpio, Pisces, Taurus	First & Second Q
Potatoes	Cancer, Scorpio, Pisces, Libra, Capricorn	Third Q
Pumpkins	Cancer, Scorpio, Pisces, Taurus	Second Q
Rye	Cancer, Scorpio, Pisces, Libra, Capricorn	First & Second Q
Oats	Cancer, Scorpio, Pisces, Libra, Capricorn	First & Second Q
Spinach	Cancer, Scorpio, Pisces, Taurus	First & Second Q
Squash	Cancer, Scorpio, Pisces, Taurus	Second Q
Sunflowers	Cancer, Scorpio, Pisces, Taurus, Libra, Capricorn	First & Second Q
Sweet Potatoes	Cancer, Scorpio, Pisces, Taurus, Libra	Third Q
Tomatoes	Cancer, Scorpio, Pisces, Taurus	Second Q
Turnips	Cancer, Scorpio, Pisces, Taurus	Third Q
Watermelon	Cancer, Scorpio, Pisces, Taurus	Second Q
Wheat	Cancer, Scorpio, Pisces, Libra, Capricorn	First & Second Q

*While considered a barren sign, many people plant runner and pole beans in Gemini with success.

©Katherine M. Grossman - 2006-2020

The Complete Granny Miller

The Best Time for Setting Fence Posts

Sooner or later, most smallholders or garden farmers will have occasion to erect a permanent fence. Whether you build it yourself or pay someone else to construct a fence, there is no getting around it - fencing is expensive. In fact, fencing can be one of the costliest capital improvements on a homestead or small farm. It often takes quite a bit of time to get good, sturdy fences put up. But once they're up, they will last for years with sensible maintenance.

When building fences, it's a false economy to go on the cheap. Build the best fence you can afford. But before you drive a fence post into the ground, you may want to consider some fence-building folk wisdom from our agrarian ancestors. Pay attention to the moon's cycle. The moon's monthly cycle affects how well a fence post will stay in the ground. The most suitable time for driving and setting fence posts is when the moon is waning during the 4th quarter. Fence posts can also be set on the day of the new moon.

Avoid the moon's 1st and 2nd quarters. Earth days are the most favorable. Taurus, Virgo, and Capricorn days are all good.

Avoid the water sign Cancer. Posts will loosen and rot more quickly when setin that sign.

With correct lunar timing, a well-set fence or gate post will stay in the ground. It will not shift nor heave, and the posts will remain rock-solid even after 30 or 40 years.

The Best Time to Make Sauerkraut

The moon's monthly cycle and its effect upon the earth are more important than most modern people realize. For many hundreds of years, folk wisdom has instructed us to make sauerkraut only when the moon is new and up until the 1st quarter. Cancer, Pisces, and Scorpio are the best moon days if they can be had. That's because when you make sauerkraut at that time, there is always plenty of juice in the crock, and the top layer of sauerkraut will not dry out.

A Candlemas Day Farming Proverb

"Half your wood and half your hay you must have on Candlemas Day."

That bit of folk wisdom is spot on. For traditional agrarian people, February 2nd marks the midpoint of winter in terms of weather. February 2nd is exactly the halfway point from the winter solstice to the first day of spring.

For planning purposes, farmers should be halfway through their hay and woodpile. February 2nd is popularly known as Groundhog Day, and is also known as Grundsaudaag in some parts of Pennsylvania. But did you know

The Complete Granny Miller

that Groundhog Day wasn't always celebrated on February 2nd?
In fact, February 14th used to be the traditional Groundhog Day.
Back in the days when we were colonies of England, the Julian calendar
(known as the Old Calendar) was officially replaced by the Gregorian
calendar (the New Calendar).
The change was implemented in small increments over a period of two
years. After everything was said and done, the change in calendars meant
that 11 dayswere dropped from everyone's life in September of 1752.
The change initially caused confusion for a while. It's the reason why
Christmas and Epiphany are sometimes referred to as New Christmas
(December 25th) and Old Christmas (January 6th).
February 2 (Old Calendar: February 14) is traditionally known as
Candlemas Day. It marks the end of the Christmas season. Candlemas is an
ancient Christian feast day, celebrated as the Presentation of Christ in the
Temple in Jerusalem.The day marks 40 days after the birth of Christ and
honors the just and devout Simeon, who embraced the Christ Child and
prayed:

"Lord, now lettest thou thy servant depart in peace, according to thy word:For
mine eyes have seen thy salvation,
Which thou hast prepared before the face of all people;
A light to lighten the Gentiles, and the glory of thy people Israel."
(Luke 2:22-39).

A Few Pennsylvania German Superstitions

Depending upon how you count them, my husband and I are the 4th or 5th
continuous generation of his family to live in our farmhouse. His family is
an old Pennsylvania German family (sometimes referred to as Pennsylvania
Dutch). They are one of the original families to have settled the Western
Pennsylvania frontier. My husband's early German ancestors intermarried
with the Scots Irish and Native American Shawnee. It's an interesting mix
of cultural and social traditions.
As some of you may be aware, Pennsylvania Germans are a superstitious
bunch. At one time, hex signs were common on barns, and every village had
a Pow Wow or Hex doctor. The belief in witches and witchcraft was
pervasive, and superstition infused the lives of many rural Pennsylvanians.
Every once in a while, I'll run across a Pennsylvania folk belief that is taken
as more than just a silly farm life superstition. One such belief is that a
scouring (diarrhea) neonatal calf can be cured by feeding it a whole egg. I
was first told about scouring calves and eggs almost 30 years ago by an old-
timer who had been a dairy farmer for many years.
At the time, I had a young Holstein bull calf named Webster. He was ill
with profuse watery yellow diarrhea. I asked the old dairy farmer for

guidance. According to the old farmer, I was to shove an entire raw egg down Webster's throat - eggshell and all.

That seemed rather extreme to me. But to make a long story short, Webster was at Death's door, and he was probably going to die anyway. There was nothing to lose by pushing an egg down his throat. So, I did it.

It may have been just a coincidence, but Webster recovered. He lived another two years before finally ending up in my freezer.

Over the years, I've collected many Pennsylvania German folk beliefs and superstitions. A few have merit and are based on a collective agrarian experience. Most are just plain weird and wacky superstitions. I thought you'd like to read a few of the more outlandish and peculiar ones.

- To prevent abortion in cows, hang a dog skull in the cow barn
- Never sew any clothing while the person is wearing it
- When you leave the house, don't turn back for something you forgot. You must sit down first
- A witch cannot cross a blue threshold
- If a bat flies in your house, it's a guaranteed sign that the devil is after you
- If you carry the 91st Psalm into battle, the bullets will not harm you
- Sneeze before breakfast, and you will have company that day
- If you receive money before 9 a.m., you will receive money all week. If you pay out money before 9 a.m., you will be paying out money all week
- Never wash clothes on Wednesday, bad luck will come to those who do. Cattle could die
- If you pick your teeth with a toothpick made from a lightning-struck tree, you will cure the toothache, but the tooth will decay
- Never shoot a cat. It's bad luck
- If you want large cucumbers, have a man plant them
- If you call a dog 'Water' or 'Fire', the dog cannot be bewitched

The Complete Granny Miller

- If your gun is bewitched, stick two pins on it in the form of a cross
- Egg shells should always be burned or crushed to prevent chickens from being bewitched
- With a newly married couple, the first one to go to sleep on the wedding night will be the first one to die
- Only move a beehive after the funeral leaves the house, or else the bees will die
- When a dog drags its rear end, it's a sure sign of a wedding
- Don't turn the wheels of a buggy or wagon backwards while greasing or fixing them. If you do, you'll have a breakdown, and the witches will bother you
- Don't take the cat with you on the day that you move house, it's bad luck. Come back and fetch her later
- Open the windows in a room after someone has died so the soul can leave
- Nail a toad's foot over the barn door to keep the witches out
- If you dream of milk, it means you will fall in love
- Sweep the house in the dark of the moon, and you will never be plagued with spiders or moths
- Spitting into a fire causes a toothache
- Never plant peas or beans on the same day that baking is done
- To cure founder in horses due to overfeeding of grain, pee on the hay before feeding it to the horses
- If a tree will not bear fruit, drive nails into the trunk
- Don't clean out cattle stalls or pens between Christmas and NewYear's Day. If you do, witches will bother you

The Complete Granny Miller

- The number of snowstorms during the winter is indicated by the number of days from the first snowfall to the next full moon, or to the first day of the following month
- "A fat wife and a big barn never did a man any harm"
- To drop a fork means a man will come for a visit; a knife means a woman

The Complete Granny Miller

Husband & Wife Trees

There are two very large maple trees standing side by side in my front yard near an old, shallow well. They are called Husband and Wife trees.

Old timers called them that, and you hardly ever hear the term anymore. It has gone out of fashion: like marrying for life and farming.

Of course, the expression is a folk term and an analogy taken from the natural world that was used long ago to describe a married couple's relationship.

A married man and woman are like two separate trees planted in different holes at the same time.
They are a permanent fixture in the landscape, and together they watch the passage of the years and seasons.

The trees are the same size, and one does not hinder the growth of the other. Because the trees stand so close together, they are not as subject to wind or icedamage as a single tree is.

The two together are more likely to survive adversity.
The trees grow close to one another.
But they are truly separate.

There is space enough between them for the wind and the air to pass.
From beneath their roots are entangled.
And how they are joined is hidden from the world.

The trees derive their sustenance from the same Source.
And one cannot be separated from the other without risking them both.

The Complete Granny Miller

Afterward

"We must let go of the life we have planned, so as to accept the one that is waiting for us." - Joseph Campbell

In July of 2019, my husband, Rick, and I sold our 5th-generation family farm. It was a decision that we didn't take lightly.
There wasn't going to be a 6th generation coming after us to take it over.

We were both well aware of how small family farms and homesteads often fall into disrepair as people grow older. We'd seen it happen countless times. It's sad. The decline isn't deliberate. It usually happens slowly and over many years. Time and again, we had witnessed a previously beautiful homestead or small farm fall into disrepair. Usually, it's because an older or elderly couple can no longer keep up with the hard work.
Why would it be any different for us?
There appears to be a form of blindness in older farmers and homesteaders, caused by an intense love for home, animals, land, and place. It doesn't permit people to see the decay and disrepair that slowly creeps in. A life of homesteading and small family farming requires a large dose of youth, and often more energy than older people possess. We didn't want what we had worked so hard for to decline and perish. We knew it was time to pass the reins to younger people. To persons who would love and care for the place as much as we had. So, we decided to sell the farm and move to town.
We knew that the move would be an enormous lifestyle and emotional adjustment for both of us. We were prepared for that. So long as we had each other, we could face anything together. We bought a house on a double-sized lot in a small, quiet Western Pennsylvania college town. We planned to make a vegetable garden and try mini-homesteading, old people style.
 Rick planned to cut back on his consulting business. We were going to travel and do things that we had never been able to do before. Aside from visiting family, Rick and I had never taken a vacation or trip for just the two of us. We planned to travel west and see the Grand Canyon, the Rockies, and the Pacific Ocean. We planned to visit Europe and Australia. Maybe even visit Greece and Jerusalem. I wanted to go to Nova Scotia and Western Canada. He wanted to go to Poland and Czechoslovakia. We looked forward to the next phase of our life together. We joked about becoming town people and growing old and fussy together.

The Complete Granny Miller

What we didn't plan for was that within 10 days of buying our new house in town, Rick would be diagnosed with stage 4 esophageal cancer. Or that he'd be dead within 10 months.

As I prepare and ready this book for re-publication, I've been a widow for 70 days. Due to COVID-19, I buried my husband alone without my children and family by my side. The country is still in quarantine and lockdown. People are uncertain and fearful about what the future holds. Meat is scarce, and toilet paper is expensive and a national joke. Seed racks at the local feed store have been nearly cleaned out, and 30 million people are unemployed. There is rioting and civil unrest all across the country due to alleged police misconduct. I'm not sure where the country is headed. I'm not exactly sure where I'm headed. Only time will reveal that.
However, what I am sure of is that my life was changed for the better because I was a homesteader and a farm wife. I understand self-reliance and personal freedom in a way that would not have been possible without that life experience. I'm not dependent or vulnerable in the same way that I was 35 years ago. I'm secure in the knowledge that the practical skills and life outlook I now possess will see me through my old age.
I plan to put in a small garden this summer and do some home canning. I will put aside food and build up my pantry for the winter months. I will find an alternative source of heat for my home, and I plan to keep some rabbits and illegal chickens in the garage if things really get crazy.
Most of all, I plan to accept whatever life has planned for me. I have absolute confidence that better days are ahead.

Katherine Grossman
Spring 2020

I sincerely hope that you have enjoyed this book and have found something useful in it for your life.
If you would like to contact me, I can be reached by email at:

grannymiller@zoominternet.net

or

Katherine Grossman
P.O. Box 116
Harrisville, Pennsylvania 16038

Author website:
www.grannymiller.com

The Complete Granny Miller

www.ingramcontent.com/pod-product-compliance
Lightning Source LLC
Chambersburg PA
CBHW031546040426
42452CB00006B/200